DIETARY LIPIDS AND POSTNATAL DEVELOPMENT

Dietary Lipids
and
Postnatal
Development

EDITORS

C. Galli
*Institute of Pharmacology
 and Pharmacognosy
University of Milan
Milan, Italy*

G. Jacini
*Stazione Sperimentale
 Olii e Grassi
Milan, Italy*

A. Pecile
*Institute of Pharmacology
 and Pharmacognosy
University of Milan
Milan, Italy*

Raven Press, Publishers · New York

Preface

This volume is a multidisciplinary account of the problems related to nutritional requirements and the biological roles of dietary lipids during early stages of postnatal development in mammals and especially in man. The relevance of lipids and lipid-soluble vitamins to growth emerges from studies carried out with several experimental models and from scattered clinical observations which are reviewed herein.

The volume is divided into five sections, each devoted to a specific aspect of the general problem. Since the physiological vehicle of dietary lipids in early postnatal development in mammals is milk, a sizable portion of the book (Section A) includes chapters which analyze the chemical and physico-chemical characteristics of milk lipids. The problems related to the replacement of these lipids in milk substitutes and the interrelationships between maternal dietary habits and content of polyenoic fatty acids in milk lipids at various stages of lactation are also discussed.

The papers in Sections B and C give some measure of the ever-growing interest in the mechanisms of lipid absorption, the enzymatic processes involved in their utilization by various tissues, the interaction with hormonal factors during tissue growth, and the specific roles of essential polyunsaturated fatty acids in various physiological and biochemical functions.

Section D is specifically devoted to studies of the effects of dietary lipids on brain development. The high rate of utilization of polyunsaturated essential fatty acids by the growing CNS suggests a distinctive requirement for these compounds in the very early stages of postnatal development. Thus, special emphasis is placed on the importance of the availability of essential fatty acids in adequate form and amounts for optimal growth of the suckling child in commercially prepared milk formulas.

The last section (Section E) deals with the problem of drug secretion into milk. The mechanisms regulating this process are considered also in the light of data obtained by the most up-to-date, highly sensitive techniques for detection of drugs and their metabolites in milk. The determination of compounds secreted into milk can be a guide to identifying those pharmacological treatments which should be avoided during lactation.

To anyone interested in the problem of postnatal development from a nutritional and a biochemical standpoint the book presents a comprehensive view of the multiple aspects concerning the biological importance of dietary lipids, which is often overlooked, especially by practitioners. We believe the volume is a valuable guide to pediatricians and nutritionists, and that it will provide investigators in the field with stimulating suggestions for further research.

The Editors

Contents

C. SELECTED TOPICS ON DIETARY LIPIDS, LIPID SOLUBLE VITAMINS AND GROWTH HORMONES DURING DEVELOPMENT

D. EFFECTS OF DIETARY LIPIDS ON THE CENTRAL NERVOUS SYSTEM

E. DRUG SECRETION INTO MILK

Dietary Lipids and Postnatal Development
Raven Press, New York © 1973

Nonglyceride Constituents of the Lipid Fraction of Milk

S. Kuzdzal-Savoie, W. Kuzdzal, D. Langlois, and J. Trehin

Station Centrale de Recherches Laitières et de Technologie des Produits Animaux, I.N.R.A., Jouy-en-Josas, FRANCE

I. INTRODUCTION

Triglycerides represent about 98% of the lipids in milk. The nonglyceride fraction that we have studied is therefore very small, but the biological role of its components appears to be more and more important. By studying the composition (1) of this nonglyceride fraction, we shall examine successively the nonlipid components of milk fat (except the liposoluble vitamins), the nonglyceride single lipids, and the complex lipids. A survey of present knowledge on biosynthesis of complex milk lipids will also be given. Finally, some aspects of the biological role of complex lipids will be reported.

II. STUDY OF THE COMPOSITION OF THE NONGLYCERIDE FRACTION

A. Nonlipid Components

1. *Natural hydrocarbons constitute about 0.1% of the milk fat and are present in the milk of all the species studied.* The milk fat of the cow (Fig. 1) contains saturated hydrocarbons with 18 to 32 or 33 carbon atoms (our results differ from those of Ristow and Werner (2a). The hydrocarbons with odd and even numbers of carbon atoms exist in equivalent proportions, but the amount of branched and unsaturated hydrocarbons is much lower. Ewe's milk contains more octadecane and octadecene. The natural hydrocarbons of woman's milk seem to be characterized by the presence of short-chain hydrocarbons (less than 20 carbon atoms) and squalene (2b) (1 mg/100 g = 10 ppm). β-Carotene represents about 0.001% of the milk fat of the cow or the woman, whereas that of the goat, ewe, and buffalo cow does not contain any appreciable amounts. There are no natural aromatic hydrocarbons in the milk.

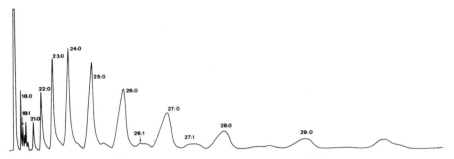

FIG. 1. Natural hydrocarbons of cow's milk fat.

2. *Cholesterol.* Woman's milk contains about 500 mg cholesterol/100 g fat versus about 300 mg/100 g fat in cow's milk. The cholesterol level is not subject to any seasonal variations (3).

3. *Alcohols.* Milk fat contains a small amount of aliphatic and terpenic alcohols (about 0.03%). In the milk of the cow or the ewe, octadecanol constitutes the major fraction, but there is also some eicosanol. Terpenic alcohols are also present, but in lower quantity.

A large proportion of the alcohols, in particular the aliphatic ones, are present in the milk in esterified form (Fig. 2).

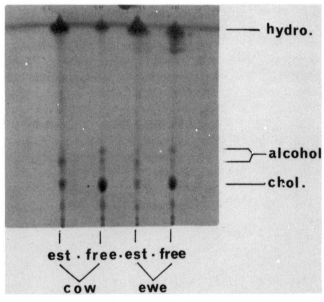

FIG. 2. Cholesterol and alcohols repartition between free and esterified forms.

B. Single Nonglyceride Lipids

The sterol esters are among the single nonglyceride lipids found in milk. Esterified cholesterol probably represents a slightly higher proportion than the 10% of total cholesterol generally admitted. Keenan and Patton (4) have studied sterol esters in the milk of the cow and the goat. We have attempted to analyze the fatty acids of sterol esters in the milk of the woman, the cow, and the ewe (Fig. 3).

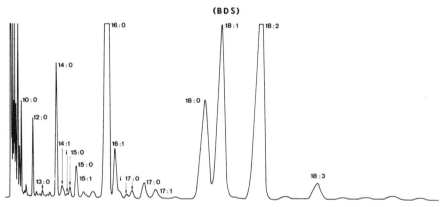

FIG. 3. Sterol ester fatty acids from ewe milk.

According to primary approximation, the fatty acid composition of sterol esters seems to be similar to that of the triglycerides, but differences which seem to be rather characteristic for the fatty acids of sterol esters have been determined; for example, we found the presence of a higher proportion of minor acids than that found in the triglycerides of corresponding milk samples. In addition, a very high level of linoleic acid in the sterol esters of ewe's milk was observed, as well as the absence of a high proportion of capric acid, which is characteristic for triglycerides. The high level of lauric acid, also characteristic for the fatty acids of triglycerides, was not found either in the sterol esters of human colostrum (Fig. 4) or in those of woman's milk. On the contrary, lauric acid is preponderant in the free fatty acids of woman's milk (Fig. 5).

The composition of the free fatty acids of raw cow's milk (about 0.1% of total acids) is similar to that of the fatty acids of the triglycerides. The level of free fatty acids generally increases at the end of lactation. Their origin is controversial, but it is likely that they proceed from an incomplete synthesis rather than from a hydrolysis prior to lactation.

Woman's milk is very rapidly hydrolyzed even when stored at low temperatures.

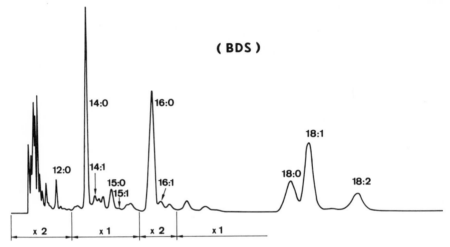

FIG. 4. Sterol ester fatty acids from human colostrum.

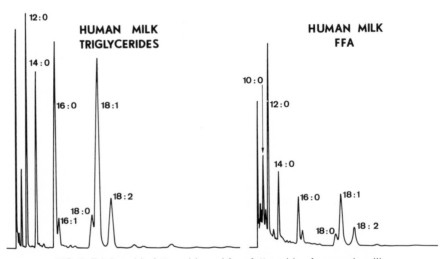

FIG. 5. Triglyceride fatty acids and free fatty acids of woman's milk.

C. Complex Lipids

The complex milk lipids constitute a very diversified group comprising a polar group associated with a fatty acid. This polar group is of varied nature: amino acid, phosphoric acid, carbohydrate, sphingosine, and so on.

1. *Level and distribution of complex milk lipids.* The complex milk lipids consist mainly of phosphoamino lipids. Cow's milk contains 0.3 to 0.4 g of phospholipids per liter, that is, about 1% of total lipids. This level falls to around 0.2% in butter and almost disappears in butter oil. Woman's milk and ass's milk contain about two times more phospholipids than cow's milk. The colostrum level of phospholipids is the same as the milk level.

Table 1 shows the distribution of complex lipids during the course of milk transformation. Skim milk and buttermilk contain the highest proportion of complex lipids compared to single lipids. The ratio of phospholipids to total lipids often reaches higher values than those indicated in the table for skim milk and buttermilk powder.

TABLE 1. *Phospholipid content (%) of milk and milk products*

Product	Phospholipids in product	Fat in product	Phospholipids in fat
Whole milk	0.0337	3.88	0.869
Skim milk	0.0169	0.090	17.29
Cream	0.1816	41.13	0.442
Buttermilk	0.1819	1.94	9.378
Butter	0.1872	84.8	0.2207

2. *Classification of polar or complex lipids and total composition.* Like any biological material, the complex milk lipids can be divided into two main classes: polar lipids containing glycerol and polar lipids containing sphingosine (or analogous bases). Each of these classes can then be subdivided into phosphorous-containing polar lipids and polar lipids without phosphorous. The glycerol- and phosphorous-containing polar lipids represent the major fraction in milk. Phosphatidyl ethanolamine, phosphatidyl choline, phosphatidyl serine, and phosphatidyl inositol belong to this group. These components may exist in three forms: diacyl, vinyl ether (called plasmalogenes), and alkoxy ether. In milk, the diacyl group is the preponder-

ant fraction. Only the aldehydogenic forms of phosphatidyl ethanolamine and phosphatidyl choline have been isolated.

The glycerol-containing polar components without phosphorous (such as the galactosyl glycerides well known in the plants used in dairy cow feeding) have not been detected in the milk. There also does not seem to be any phosphatidic acid or phosphatidyl glycerol, well-known components in the complex lipids of the brain, the heart, and human blood serum (5), nor do we find any cardiolipids, even though this component is present in the lipids of the mammary tissue (6).

The second category of polar lipids consists of components containing sphingosine or a similar amine base. Grouped under the name of sphingolipids, they are subdivided into phosphorated sphingolipids and nonphosphorated sphingolipids. Sphingomyelin, the major polar lipid of milk (20% of total polar lipids), belongs to the first group. Sphingomyelin contains choline and is characterized by an amide function (it can be called ceramide phosphorylcholine). In the second group, without phosphorous, we find the ceramides, proceeding from the combination of sphingosine and a fatty acid. The milk contains mono- and dihexoside ceramides (glucosylceramide, 40%; lactosylceramide, 60%). These components may be compared with cerebrosides, but according to Kayser and Patton (7) the difference from the latter is that they do not contain any hydroxylated acids, whereas Morrison and Hay (8) find that these represent 1% of the acids.

Table 2 shows the relative proportions of the various polar lipids in cow's milk. In fact, in the milk of all species we find proportions similar to those reported in this table. This is based on the findings of Morrison (9), who analyzed the polar lipids in the milk of seven mammalian species. The polar lipids of woman's milk contain a little less phosphatidyl ethanolamine and phosphatidyl choline than cow's milk, but more sphingomyelin and phosphatidyl serine.

TABLE 2. *Mean composition of cow's milk polar lipids*

Name	Percent of total weight
Ceramide monohexoside	3
Ceramide dihexoside	3
Diacylphosphatidyl ethanolamine	30
Vinyl ether phosphatidyl ethanolamine	1
Phosphatidyl serine	8
Phosphatidyl inositol	5
Diacyl phosphatidyl choline	28
Vinyl ether phosphatidyl choline	3
Sphingomyelin	19

The monoacylated components (lysophosphatidyl ethanolamine and lysophosphatidyl choline) are also present in woman's milk at a higher level (5% of total polar lipids) than in cow's milk (scarcely 1% of total polar lipids). In all milk samples studied, the proportion of choline-containing phospholipids is constant and close to 60% of total polar lipids (9).

3. *Study of the fatty acids of polar lipids.* The distribution scheme of the fatty acids of polar lipids in milk is different from that characterizing neutral lipids (Fig. 6). Even though there is a difference between woman's milk

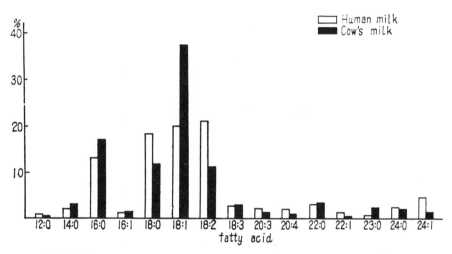

FIG. 6. Phospholipid fatty acids from cow's milk fat and woman's milk fat.

and cow's milk as regards this scheme, it is always characterized by the absence of short-chain fatty acids, a higher proportion of linoleic acid, and a notable proportion of long-chain fatty acids (generally indicated for neutral lipids, but at a level 20 to 30 times lower).

As regards these long-chain fatty acids, there are some differences between woman's milk and cow's milk: more nervonic acid (24:1) in the phospholipids of woman's milk, more 23:0 in the phospholipids of cow's milk. There is only a small quantity of arachidonic acid, which is generally a major constituent of the phospholipids of animal tissues.

Fractionation of milk phospholipids allows a more thorough study of the different milk polar lipids. In addition, some authors have determined the nature of the fatty acids present in the position α' or β of the glyceride chain.

Phosphatidyl ethanolamine is the most unsaturated phospholipid fraction (Table 3). The proportion of oleic acid is high in the milk of all species studied. It is lower in woman's milk, however, where the high proportion

TABLE 3. *Partial fatty acid pattern of milk PE*
(% of total PE fatty acids)

	Human	Cow	Ewe	Sow	Camel	Buffalo	Ass
18:1 (oleic acid)	16	50	50	35	26	40	35
18:2 (linoleic acid)	18	12	8	18	18	13	23
20:4 (arachidonic acid)	12.5	1	1	6.6	?	?	1

of linoleic and arachidonic acid provides a compensatory unsaturation. The relative proportions of these two polyunsaturated acids are also high in the phosphatidyl ethanolamine of cow's milk (18 and 6%). There is only a small proportion of arachidonic acid (0 to 1%) in the phosphatidyl ethanolamine of cow's, ewe's, and ass's milk. The level of linoleic acid remains high (23%) in the phosphatidyl ethanolamine of ass's milk (10–12). All arachidonic acid present in phosphatidyl ethanolamine and 90% of the linoleic acid are located in position 2 or β of sn-glycerol-3 phosphate. Oleic acid is distributed in positions 1 and 2. Stearic and palmitic acids are located in external positions.

Phosphatidyl choline is more saturated than phosphatidyl ethanolamine. Palmitic, stearic, and oleic acids are the major fatty acids of the latter. There is also a small proportion of linoleic acid (5 to 10%). Recently, Hay and Morrison reported that the phytanic acid (1% of the fatty acids) of phosphatidyl choline occurred only in position 1(13). The distribution of fatty acids according to the internal or external positions of the glyceride chain is rather similar to that determined for phosphatidyl ethanolamine, that is, saturated external position and unsaturated internal position.

The fatty acid composition of phosphatidyl serine and phosphatidyl inositol of the different milk samples has also been determined, although the separation of these two fractions is difficult. The fatty acid composition is similar to that of phosphatidyl choline.

On the contrary, sphingosides show a clear differentiation according to animal species. In this fraction we find the long-chain fatty acids whose presence in the fatty acids of phospholipids is one of the characteristics of the difference toward neutral milk lipids. Sphingomyelin of milk is the most saturated phospholipid fraction. The proportion of saturated acids always exceeds 80%, and reaches 90 and 96%, respectively, in ewe's milk and buffalo cow's milk.

In the case of ruminants, this saturation is due to a high proportion of palmitic acid and saturated acids with 22, 23, and 24 carbon atoms. This distribution scheme is given in Fig. 7 and shows the fatty acids of sphingomyelin in woman's and cow's milk. Compared to 22:0, we find more 23:0 in sphingomyelin of cow's milk, and more nervonic acid in the sphingomyelin of woman's and ass's milk. The fatty acid composition of sphingomyelin of camel's milk is intermediate between that of woman's milk and cow's milk.

The monounsaturated fatty acids in the sphingomyelin of cow's milk have a high proportion of *trans* acids (60% of the monoenes with 23 carbon atoms, 25% of the monoenes with 24 carbon atoms). The position of the double bond is preferential in the *cis* monoenes ($\omega 9$) but varies much in the *trans*-monoenes (from $\omega 5$ to $\omega 9$). Thus, the latter would proceed from a selective elongation of the *trans*-monoene acids of the triglycerides of cow's milk.

Morrison and Hay (8) have also studied the long-chain bases of sphingolipids. There is a steric isomer on carbon 4, and ramifications are noticed on the methylene chain. Morrison has identified 31 long-chain bases in the sphingolipids of cow's milk. The dehydroxysaturated bases constitute 80% of the total bases. 4D-Sphingenine (or sphingosine), with 18 carbon atoms, and hexadecasphingenine are preponderant.

The fatty acid composition of ceramides (sphingosides without phosphorous) is similar to that of sphingomyelin, but the proportion of acids with 22, 23, and 24 carbon atoms reaches 90% of the total acids.

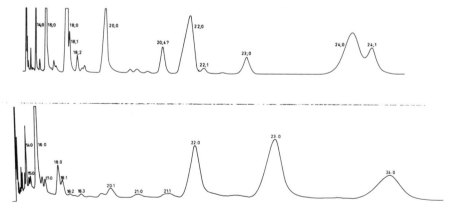

FIG. 7. Sphingomyelin fatty acids from cow's and woman's milk.

D. Diacylglyceryl Ethers

Diacylglyceryl ethers are nonglyceride components, *sensu stricto,* since they are ether-esters. They belong to the complex lipid group although they differ from the phosphatidyl glyceryl ethers studied in relation to the phospholipids. There is a high amount of diacylglyceryl ethers in the lipids of the liver or flesh of some fishes. They also exist in mammalian tissues, but in small quantities.

Milk lipids contain diacylglyceryl ethers. The amount present in woman's milk is ten times higher than that in cow's milk, that is, 0.1% of total woman's milk lipids (14). The amount present in human colostrum is a little higher. On Fig. 8, diacylglyceryl ethers appear as a weak spot just over the spot of the triglycerides.

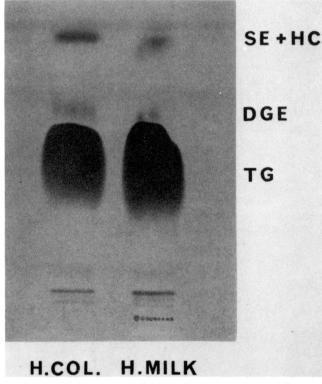

FIG. 8. Thin-layer chromatography of the total lipids of woman's milk fat and human colostrum fat.

After isolation and saponification of diacylglyceryl ethers, we obtain, on the one hand, fatty acids and, on the other hand, diols or glyceryl ethers.

In woman's milk, Halgren and Larson (14) identified, in addition to chimyl, batyl, and selachyl alcohols (80% of glyceryl ethers), saturated monoene and diene alcohols with 17, 19, 20, 22, and 24 carbon atoms. By using propylidene derivatives and gas chromatography analysis, we confirmed the presence in high proportion of chimyl, batyl, and selachyl alcohols, but we also found a large proportion of unsaturated diol with 16 carbon atoms.

Figure 9 shows the distribution scheme of the fatty acids of diacylglyceryl ether in human colostrum. The proportion of palmitic acid is high, that of lauric acid low. Consequently, the distribution pattern is similar to that of sterol esters.

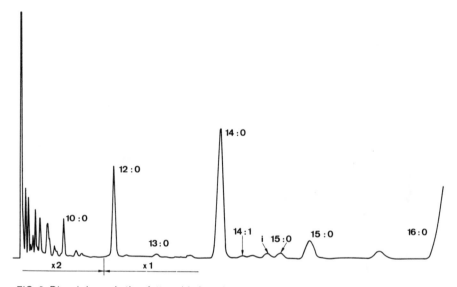

FIG. 9. Diacylglyceryl ether fatty acids from human colostrum fat (partial chromatogram).

III. BIOSYNTHESIS OF NONGLYCERIDE LIPIDS

A. Fundamental and General Aspect

1. *Cholesterol.* The cholesterol of milk has a double origin: on the one hand, it is synthesized by the mammary gland *in situ* from acetate and, on the other hand, the mammary gland takes up part of the blood cholesterol (4, 15).

2. *Sterol esters.* Sterol esters are formed from cholesterol and fatty acids in the form of acyl-coenzyme A. According to Keenan and Patton (4), the fatty acid composition of milk sterol esters is similar to that of the sterol esters in the mammary tissue. However, even though this particularity is well established in the case of "milk" still retained in the mammary gland, it seems to be less obvious for the fatty acids of sterol esters in the normal secreted milk.

In the mammary gland, the turnover of the fatty acids of sterol esters is very rapid. These components are probably used to ensure the transfer of fatty acids from the available "pool" toward the sites of assembling or formation of glycerides. The specific composition of the fatty acids of sterol esters would therefore be the result of a preferential retention.

3. *Phospholipids.* By means of an asymmetric phosphorylation of the apparently symmetric glycerol molecule, in the presence of glycerol kinase, we obtained exclusively 3-sn-glycerophosphate. The reduction of dihydroxy acetone phosphate may also lead to 3-sn-glycerophosphate.

3-sn-Glycerophosphate is acylated in the 1 and 2 positions. This reaction is catalyzed by acyl transferase. The fatty acids involved in the reaction are previously "activated" and transformed into coenzyme A esters. Thus, we obtain 1,2-diacyl-3-sn-glycerophosphate or phosphatidic acid. During the formation of phosphatidic acids, a preferential acylation is observed. Palmitic acid and the unsaturated acids are more rapidly acylated than, for instance, stearic acid, and the synthesis is faster with a monoene-saturated, monoene-diene, or diene-saturated combinations (16). The incorporation of arachidonic acid takes place on a monoacyl phospholipid.

Acylation of fatty acids does not take place randomly during the synthesis of the phosphatidic acid, whereas the next stage, the dephosphorylation of phosphatidic acid (by phosphatidase phosphohydrolase), does not present any selectivity (16).

1,2-Diacyl-sn-glycerol is the common precursor of triglycerides and glycero phospholipids (17).

Phosphorylation of a diglyceride is also possible (in the presence of ATP and diglyceride kinase), but such a process does not intervene in the synthesis of phosphatidyl choline, since phosphoryl choline is incorporated intact into the developed lipid. Choline, in the presence of ATP and choline kinase, is transformed into phosphoryl choline, whose active form is cytidine diphosphocholine, which is essential for the transfer of phosphoryl choline on the 1,2-diglyceride derived from phosphatidic acid. The latter is therefore continuously synthesized and degraded.

In the presence of 1-alkenyl-2-acylglycerol, we obtain a plasmalogene after addition of the phosphorylated base.

By means of decarboxylation, serine leads to ethanolamine which, by methylation, forms choline. Serine is the precursor of ethanolamine. Sphingosine proceeds from the condensation of palmitic acid and serine.

There are two possible pathways for the formation of sphingomyelin. In the first, sphingosine is acylated. The developed ceramide is used as acceptor of the phosphoryl choline group. In the second pathway, phosphoryl choline is transferred on the free sphingosine and the acylation only occurs afterward. Consequently, there is a bond between a phosphorated base and the diglycerid or between a phosphatidic acid and the base.

Synthesis of phosphatidyl inositol supposes formation of activated phosphatidic acid which is bound to the free inositol. Synthesis of phosphatidyl serine and phosphatidyl ethanolamine also takes place through the diglyceride pathway (17).

A choice is observed among the available diglycerides at the moment of the definitive synthesis. There is a preference for diglycerides rich in hexaenes in the case of formation of phosphatidyl ethanolamine and diglycerides rich in tetraenes for the formation of phosphatidyl choline (16).

The specificity toward fatty acids at the synthesis of phosphatidic acid is sufficient to explain the differences which separate the two distribution schemes of the fatty acids: on the one hand, triglycerides and, on the other hand, glycerophospholipids (18).

Synthesis of cerebrosides takes place in two ways: either acylation of a long-chain base followed by galactosylation, or galactosylation of a long-chain base followed by acylation. The presence of hydroxylated acids facilitates the acylation (19).

4. *Synthesis of diacylglyceryl ethers.* In neutral lipids, the biosynthesis of an ether bond proceeds from a condensation between a fatty alcohol and the carbonyl group of dihydroxyacetone phosphate. The vinyl ether group of a plasmalogene results from a biodehydrogenation of the corresponding alkyl-ether (hydroxylation followed by dehydration) (20–22).

B. Particular Aspects of the Biosynthesis of Complex Milk Lipids

The differences between milk triglycerides and phospholipids with respect to fatty acid composition and specific distribution of the fatty acids exclude their direct biosynthesis from a common precursor, diglyceride or phosphatidyl serine (23). In addition, milk phospholipids differ considerably from blood phospholipids (24). Phospholipids in the mammary tissue of dairy cows and those of the milk have similar global and fatty acid compositions (6). Therefore, the mammary gland is the active site for the synthesis of complex milk lipids.

The mammary gland uses a single phosphate "pool" inside the tissues and provides both the inorganic phosphorous of the milk and the phosphorous of the phospholipids in the milk. The phosphorous of the phospholipids in the blood serum is not used to form phospholipids in the milk, nor are phospholipids used that are consumed. However, comparison (Fig. 10) between the analysis of complex lipids in the mammary tissue and of complex lipids in the milk shows that some components of the mammary gland lipids do not exist in the milk lipids. On the contrary, high amounts of some other components are present in the complex lipids of the milk. The latter contain more sphingomyelin, more phosphatidyl ethanolamine, more ceramides, less phosphatidyl serine, and less inositol than the phospholipids of the mammary tissue.

In fact, milk phospholipids do not represent total phospholipids of the mammary tissue. Cardiolipid, for instance, which is present in the lipids of the mammary tissue, does not exist in the milk lipids. This component is characteristic for phospholipids of mitochondrial membranes (26, 27), and is present in the mitochondrial membranes of the bovine mammary tissue (28). Consequently, during the secretion of fatty globules there is no passage of mitochondrial components in the milk.

The composition of the phospholipids in the membranes of the different constituents of the secretory cell has been determined. Thus, the composition of the phospholipids in the microsomes of the mammary tissue has been studied (29). It differs from the composition of the phospholipids present

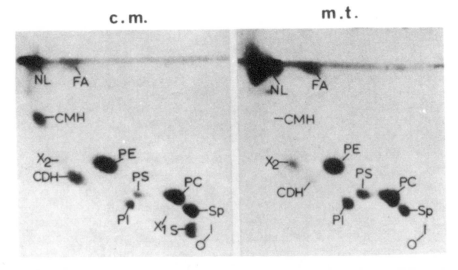

FIG. 10. Bidimensional thin-layer chromatography of cow's milk (c.m.) polar lipids and mammary tissue (m.t.) polar lipids.

in the membrane of the fatty globules of the milk. The microsomes of the mammary tissue, which contain more phosphatidyl ethanolamine and less sphingomyelin than the milk phospholipid, are not the direct source for the phospholipids of the fatty globule membrane, as it was imagined years ago (30). On the other hand, microsomes of cow's milk proceed from the microsomes of the cell (31).

The phospholipids of the cytoplasmic membrane (32–36) were analyzed in the same way. Sphingomyelin represents 33%, and ceramides 20%, of the total phospholipids. The contribution of phospholipids from the cytoplasmic membrane to the formation of milk phospholipids thus explains the above-noticed differences between milk phospholipids and total phospholipids of the mammary tissue (Fig. 10).

Observations made by means of electron microscopy also confirm the results of the chemical analyses. Thus, membrane fragments have, for example, been detected in skim milk (37). The existence of a great number of corpuscles or vesicles has been observed at the periphery of the fatty globule still enclosed in the cytoplasma of the cell, but having reached its apex. At present, they are supposed to be the membranes of the Golgi corpora (38).

Consequently, the complex lipids in the membrane of the milk fatty globules originate from the complex lipids of the plasma membrane of the cell, and the complex lipids in the lipoproteins of skim milk proceed from the membranes of the Golgi corpora of the cell. The phospholipids of skim milk lipoproteins show the same composition and the same fatty acid distribution scheme as the phospholipids in the membrane of the fatty globules, but do not arise from the disintegration of this membrane (34).

A certain number of arguments support this conception of the origin of milk phospholipids. The proteins of the fatty globule membrane are identical to those of the plasma membrane (36). The enzymes present in the membrane of woman's milk fatty globules are identical to those of the plasma membrane (39). 5'-Nucleotidase, present in the plasma membranes of the cell, also exists in the membrane of the fatty globules. In addition, the presence of UDP galactosyl transferase in skim milk also supports the role played by phospholipids in the membranes of the Golgi corpora (34). In fact, this enzyme is located in the Golgi apparatus of the mammary tissue, for which it constitutes a "marker."

IV. Biological Role of Complex Lipids

Phospholipids are essential constituents of all cellular membranes. The distribution scheme of the different complex lipids is specific for the various membranes.

Fasting or feeding with poor fat content does not change the level of phospholipids in the tissues (41). Changes in the repartition of fatty acids can therefore happen (42). The cholesterol/phospholipid ratio is also definite in the membranes.

These different compositions probably correspond to different functions. In the case of abnormal function of the cells, the differentiation of the phospholipids of the various membranes disappears: the composition of the phospholipids of these membranes tends to adopt the same pattern (43).

Parallel to the specific composition of the phospholipids, fatty acids also show a specific distribution scheme. However, it has been noticed that the fatty acid composition of milk phospholipids varies within certain limits. For example, according to Saito et al. (44), the nervonic acid level of woman's milk varies from 3 to 7% of the fatty acids of the phospholipids. Is this an essential point? Probably not. It may be a nutritional aspect, but it is not directly concerned since the fatty acid level provided from milk phospholipids is relatively low and may have a negligible effect on the supply of polyunsaturated acids. On the other hand, the phospholipids of milk are not used as such by the organism. The role of these lipids is probably mainly a physicochemical one, that is, maintenance of the colloidal stability, protection of lipids against alterations caused by enzymes, and so on. However, the complex lipids at their arrival into the milk are accompanied by a great number of elements of cellular origin. The properties of this cell fraction, its enzymatic potential included, are perhaps so important that 1% complex lipids are more essential than 99% glyceride lipids.

REFERENCES

1. Kuzdzal-Savoie, S. (1971): *Ann. Nutr. Alim,* 25:A225.
2a. Ristow, R. von, and Werner, H. (1968): *Feite Seifen Anstrich.,* 70:273.
2b. Bracco, V., Hidalgo, J., and Bohren, H. (1972): *J. Dairy Sci.,* 55:165.
3. Herder, P. C. den, and Kruisheer, G. L. (1953): *XIII Cong. Lait. La Haye,* 2(4):1359.
4. Keena, T. W., and Patton, S. (1970): *Lipids,* 5:42.
5. Schwarz, H. P., and Dreisbach, L. (1970): *Biochim. Biophys. Acta,* 210:436.
6. Parsons, J. G., and Patton, S. (1967): *J. Lipid Res.,* 8:696.
7. Kayser, S. G., and Patton, S. (1970): *Biochem. Biophys. Res. Comm.,* 41:1572.
8. Morrison, W. R., and Hay, J. D. (1970): *Biochem. Biophys. Acta,* 202:460.
9. Morrison, W. R. (1968): *Lipids,* 3:101.
10. Morrison, W. R., Jack, E. L., and Smith, L. M. (1965): *J. Am. Oil Chem. Soc.,* 42:1142.
11. Morrison, W. R. (1967): *Lipids,* 2:178.
12. Morrison, W. R. (1968): *Lipids,* 3:107.
13. Hay, J. D. (1971): *Biochem. Biophys. Acta,* 248:71.
14. Halgren, B., and Larson, S. (1962): *J. Lipid Res.,* 3:39.
15. Nes, W. R. (1971): *Lipids,* 6:219.
16. Akesson, B., Elovson, J., and Arvidson, G. (1970): *Biochim. Biophys. Acta,* 210:15.
17. Gurr, M. I., and James, A. T. (1970): *Lipid Biochemistry.* Chapman & Hall, London.
18. Christie, W. W., and Moore, J. H. (1970): *Biochim. Biophys. Acta,* 210:46.
19. Kanfer, J. N., and Sargent, A. (1971): *Lipids,* 6:682.

20. Blank, M. L., Wykle, R. L., Piantadosi, C., and Snyder, F. (1970): *Biochim. Biophys. Acta*, 210:442.
21. Thompson, G. A., Jr. (1968): *Biochim. Biophys. Acta*, 152:409.
22. Snyder, J. (1969): *Progress in the Chemistry of Fats and Other Lipids*, Vol. X. p. 319.
23. Hawke, J. C. (1963): *J. Lipid Res.*, 4:255.
24. Patton, S., and McCarthy, R. D. (1963): *J. Dairy Sci.*, 46:916.
25. Easter, D. J., Patton, S., and McCarthy, R. D. (1971): *Lipids*, 6:844.
26. Chan, S. K., and Lester, R. L. (1970): *Biochim. Biophys. Acta*, 210:180.
27. Courtade, S. A., and McKibbin, J. M. (1971): *Lipids*, 6:260.
28. Huang, C. M., and Keenan, T. W. (1971): *J. Dairy Sci.*, 54:1395.
29. Kinsella, J. E. (1972): *Lipids*, 7:165.
30. Morton, R. K. (1954): *Biochem. J.*, 57:231.
31. Morton, R. K. (1953): *Nature*, 171:734.
32. Kinsella, J. E. (1971): *J. Dairy Sci.*, 54:1014.
33. Patton, S. (1966): *J. Dairy Sci.*, 49:737.
34. Patton, S., and Keenan, T. W. (1971): *Lipids*, 6:58.
35. Keenan, T. W., Olson, D. E., and Mollenhauer, H. H. (1971): *J. Dairy Sci.*, 54:295.
36. Keenan, T. W., Marre, D. J., Olson, D. E., Yunghans, W. N., and Patton, S. (1970): *J. Cell Biol.*, 44:80.
37. Stewart, P. S., Puppione, D. L., and Patton, S. (1972): *Z. Zellforsch. snikrosk. Anat.*, 123:161.
38. Wooding, F. P. B. (1971): *J. Cell Sci.*, 9:805.
39. Martel-Pradel, M. B., and Got, R. (1972): *FEBS Letters*, 21:220.
40. Huang, C. M., and Keenan, T. W. (1972): *J. Dairy Sci.*, 55:862.
41. Hill, J. G., Kuksis, A., and Beveridge, J. M. R. (1964): *J. Am. Oil Chem. Soc.*, 42:383.
42. Hill, J. G., Kuksis, A., and Beveridge, J. M. R. (1964): *J. Am. Oil Chem. Soc.*, 42:137.
43. Bergelson, L. D., Dyatlovitskaya, E. V., Torkhovskaya, T. I., Sorokina, I. B., and Gorkova, N. P. (1970): *Biochim. Biophys. Acta*, 210:287.
44. Saito, K., Furnichi, E., Kondo, S., Kavanishi, G., Nishikawa, I., Nakazato, H., Noguchi, Y., Doi, T., Noguchi, A., and Shingo, S. (1965): *Tokio Snow Brand Milk Products Co. Ltd.*, *Dairy Sci. Abstr.* 27(no. 3263).

APPENDIX

STRUCTURAL FORMULAS OF SOME COMMON LIPIDS

$$H_3C-(CH_2)_X -C-O-CH_2$$

Cardiolipin

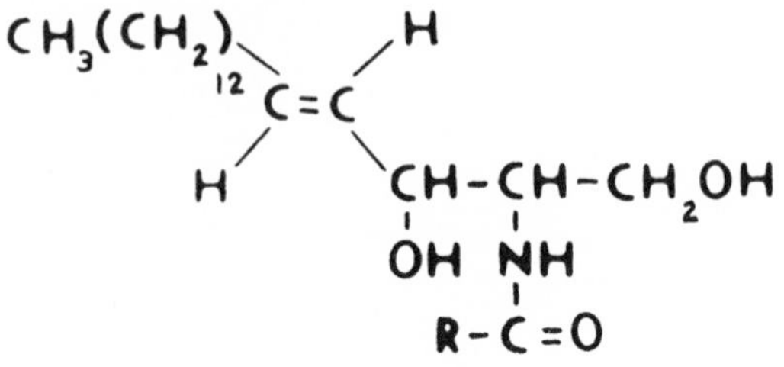

$$CH_3(CH_2)_{12} \underset{H}{\overset{}{C}} = \underset{\underset{\underset{R-C=O}{NH}}{\underset{OH}{CH-CH-CH_2OH}}}{\overset{H}{C}}$$

CERAMIDE

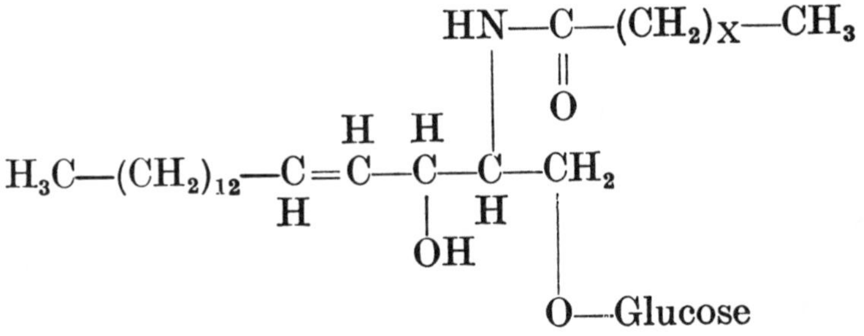

$$
\begin{array}{c}
HN-\overset{\overset{\textstyle O}{\|}}{C}-(CH_2)_X-CH_3 \\
H_3C-(CH_2)_{12}-\underset{H}{\overset{H}{C}}=\underset{\underset{OH}{|}}{\overset{H}{C}}-\underset{}{C}-\underset{H}{\overset{}{C}}-CH_2 \\
O-\text{Glucose}
\end{array}
$$

Cerebroside

$$
\begin{array}{c}
HN-\overset{\overset{\textstyle O}{\|}}{C}-(CH_2)_X-CH_3 \\
H_3C-(CH_2)_{12}-\underset{H}{\overset{H}{C}}=\underset{\underset{OH}{|}}{\overset{H}{C}}-\underset{}{C}-\underset{H}{\overset{}{C}}-CH_2
\end{array}
$$

Galactose
|
Glucose—Neuraminic Acid
|
Hexosamine

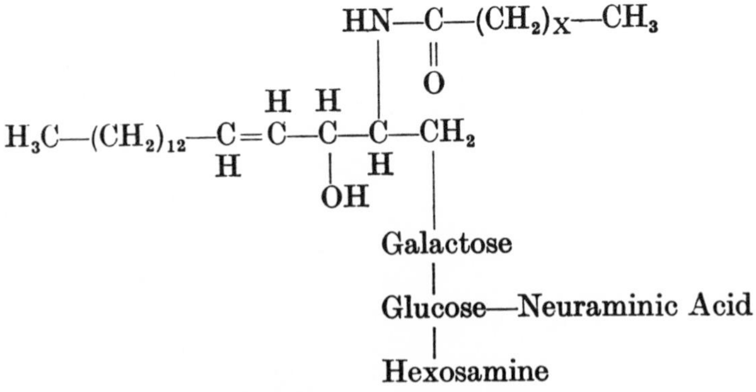

Ganglioside

Glycerol-Containing Neutral Lipids.

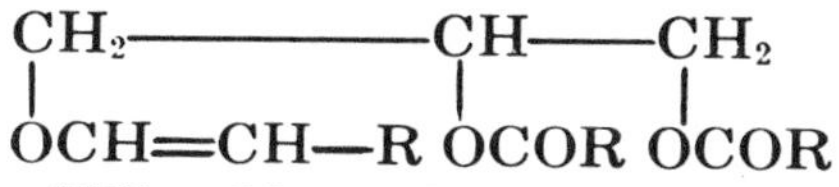

$$CH_2\!-\!\!-\!\!-\!\!-\!\!-CH\!-\!\!-\!CH_2$$
$$\underset{\displaystyle OCH\!=\!CH\!-\!R}{|}\quad \underset{\displaystyle OCOR}{|}\quad \underset{\displaystyle OCOR}{|}$$

(IX)   Neutral plasmalogen

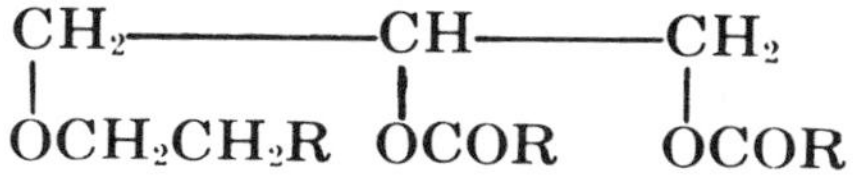

$$CH_2\!-\!\!-\!\!-\!\!-\!\!-CH\!-\!\!-\!CH_2$$
$$\underset{\displaystyle OCH_2CH_2R}{|}\quad \underset{\displaystyle OCOR}{|}\quad \underset{\displaystyle OCOR}{|}$$

(X)   Neutral glyceryl ether

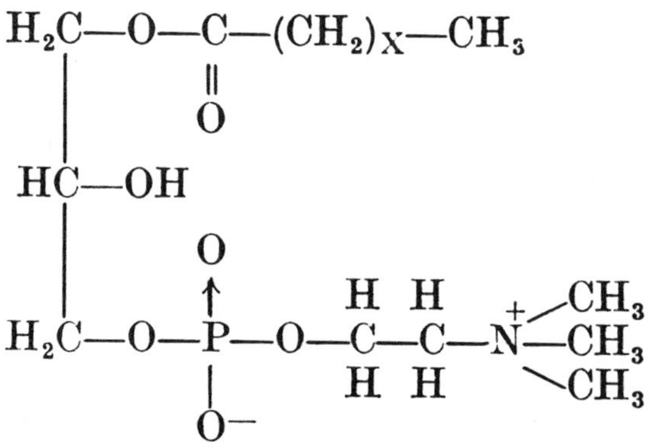

$$H_2C\!-\!O\!-\!\underset{\displaystyle O}{\overset{\displaystyle \|}{C}}\!-\!(CH_2)_X\!-\!CH_3$$

$$HC\!-\!OH$$

$$H_2C\!-\!O\!-\!\underset{\displaystyle O^-}{\overset{\displaystyle O}{\underset{\displaystyle |}{\overset{\displaystyle \uparrow}{P}}}}\!-\!O\!-\!\underset{H\ \ H}{\overset{H\ \ H}{C\!-\!C}}\!-\!\overset{+}{N}\!\!\underset{\diagdown CH_3}{\overset{\diagup CH_3}{-CH_3}}$$

Lysolecithin

$$H_2C\!-\!O\!-\!\underset{\displaystyle O}{\overset{\displaystyle \|}{C}}\!-\!(CH_2)_X\!-\!CH_3$$

$$HC\!-\!O\!-\!\underset{\displaystyle O}{\overset{\displaystyle \|}{C}}\!-\!(CH_2)_X\!-\!CH_3$$

$$H_2C\!-\!O\!-\!P\!\!\underset{\diagdown O^-}{\overset{\diagup O^-}{\rightarrow O}}$$

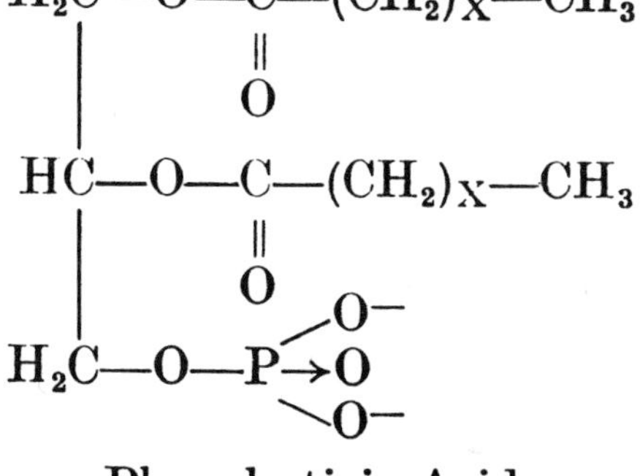

α-Phosphaticic Acid

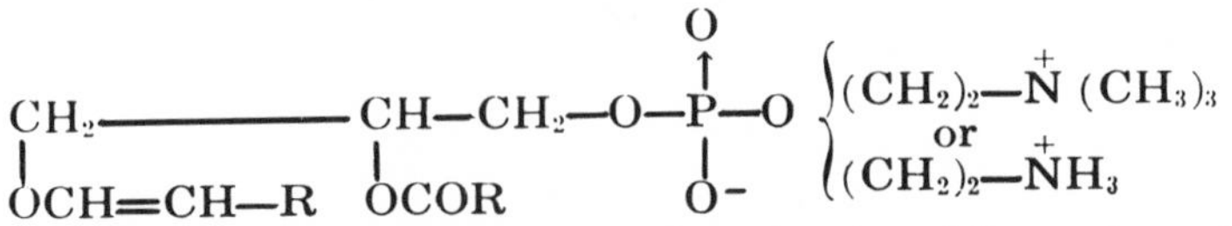

$$CH_2\text{------}CH\text{---}CH_2\text{---}O\text{---}\overset{\overset{O}{\uparrow}}{P}\text{---}O \begin{cases} (CH_2)_2\text{---}\overset{+}{N}\,(CH_3)_3 \\ \text{or} \\ (CH_2)_2\text{---}\overset{+}{N}H_3 \end{cases}$$

with $\overset{|}{O}CH{=}CH\text{---}R$ $\overset{|}{O}COR$ $\overset{|}{O}^-$

Phosphatidal choline or ethanolamine (Plasmalogen)

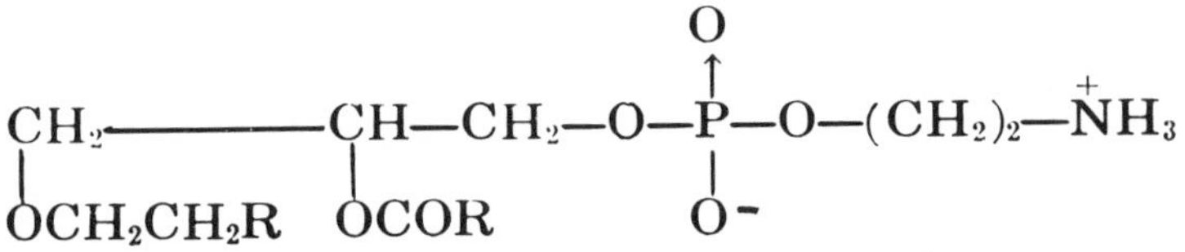

$$CH_2\text{------}CH\text{---}CH_2\text{---}O\text{---}\overset{\overset{O}{\uparrow}}{P}\text{---}O\text{---}(CH_2)_2\text{---}\overset{+}{N}H_3$$

with $\overset{|}{O}CH_2CH_2R$ $\overset{|}{O}COR$ $\overset{|}{O}^-$

Phosphatidic glyceryl ether

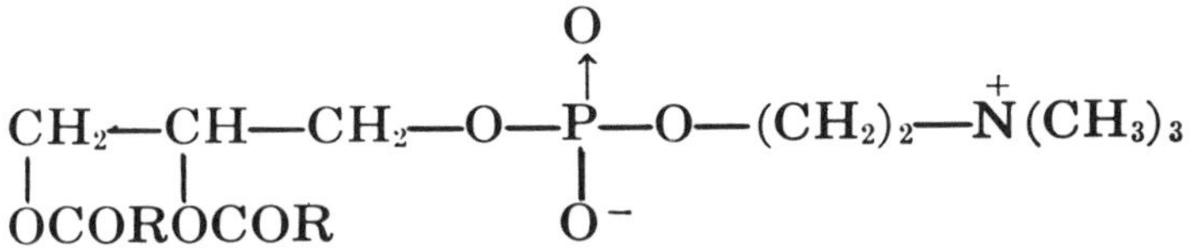

$$CH_2\text{---}CH\text{---}CH_2\text{---}O\text{---}\overset{\overset{O}{\uparrow}}{P}\text{---}O\text{---}(CH_2)_2\text{---}\overset{+}{N}(CH_3)_3$$

with $\overset{|}{O}COR\overset{|}{O}COR$ $\overset{|}{O}^-$

Phosphatidyl choline (Lecithin)

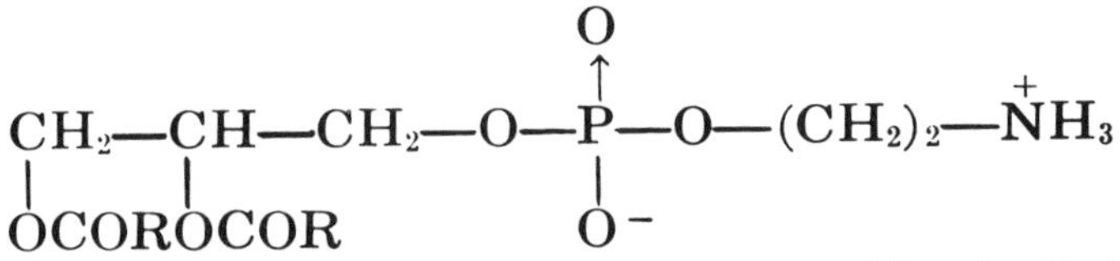

$$CH_2\text{---}CH\text{---}CH_2\text{---}O\text{---}\overset{\overset{O}{\uparrow}}{P}\text{---}O\text{---}(CH_2)_2\text{---}\overset{+}{N}H_3$$

with $\overset{|}{O}COR\overset{|}{O}COR$ $\overset{|}{O}^-$

Phosphatidyl ethanolamine (Cephalin)

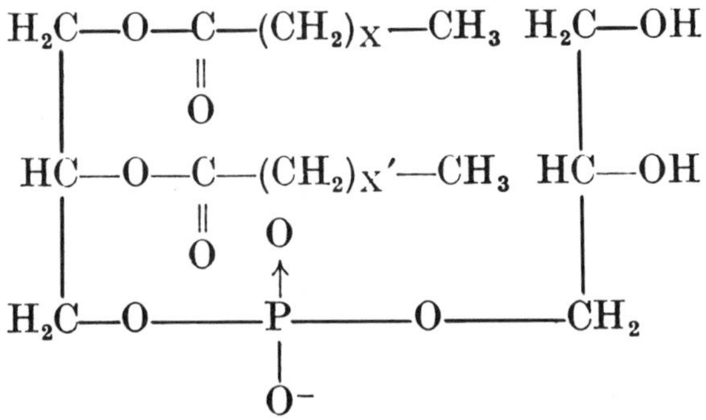

$$H_2C\text{---}O\text{---}\underset{\underset{O}{\|}}{C}\text{---}(CH_2)_X\text{---}CH_3 \quad H_2C\text{---}OH$$

$$HC\text{---}O\text{---}\underset{\underset{O}{\|}}{C}\text{---}(CH_2)_{X'}\text{---}CH_3 \quad HC\text{---}OH$$

$$H_2C\text{---}O\text{------}\underset{\underset{O^-}{|}}{\overset{\overset{O}{\uparrow}}{P}}\text{------}O\text{------}CH_2$$

Phosphatidyl Glycerol

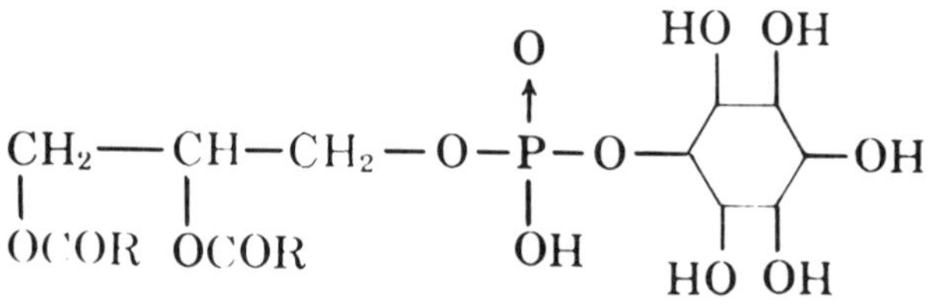

$$CH_2\!-\!CH\!-\!CH_2\!-\!O\!-\!\overset{\overset{\displaystyle O}{\uparrow}}{\underset{\underset{\displaystyle OH}{|}}{P}}\!-\!O\!-\!\text{(inositol ring: HO, OH, OH, HO, OH)}$$

OCOR OCOR

Phosphatidyl inositol

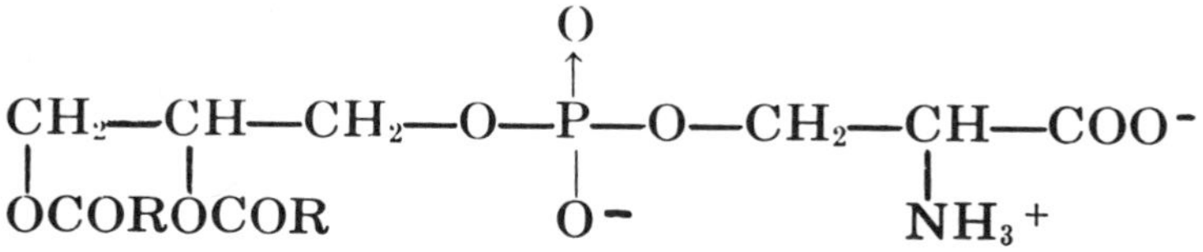

$$CH_2\!-\!CH\!-\!CH_2\!-\!O\!-\!\overset{\overset{\displaystyle O}{\uparrow}}{\underset{\underset{\displaystyle O^-}{|}}{P}}\!-\!O\!-\!CH_2\!-\!CH\!-\!COO^-$$

OCOROCOR $NH_3{}^+$

Phosphatidyl serine

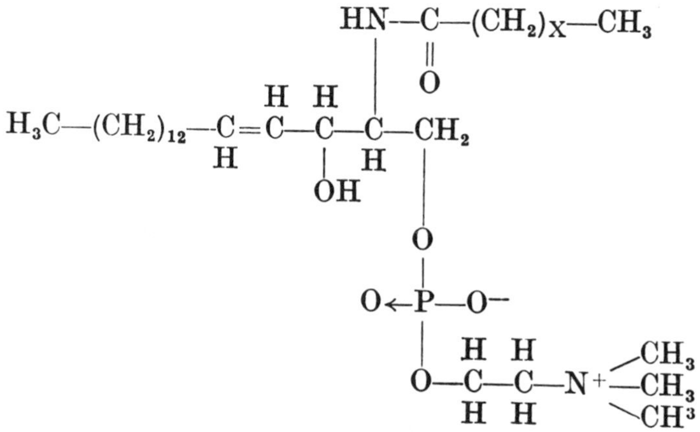

$$HN\!-\!\overset{\overset{\displaystyle O}{\|}}{C}\!-\!(CH_2)_X\!-\!CH_3$$

$$H_3C\!-\!(CH_2)_{12}\!-\!\underset{H}{C}\!=\!\underset{\underset{\displaystyle OH}{|}}{\overset{\overset{\displaystyle H}{|}}{C}}\!-\!\underset{\underset{\displaystyle H}{|}}{\overset{\overset{\displaystyle H}{|}}{C}}\!-\!CH_2$$

$$O$$
$$O\!\leftarrow\!P\!-\!O^{--}$$

$$O\!-\!\underset{\underset{\displaystyle H}{|}}{\overset{\overset{\displaystyle H}{|}}{C}}\!-\!\underset{\underset{\displaystyle H}{|}}{\overset{\overset{\displaystyle H}{|}}{C}}\!-\!N^+\!\overset{\diagup CH_3}{\underset{\diagdown CH^3}{-CH_3}}$$

Sphingomyelin

$$CH_3\!-\!(CH_2)_{12}\diagdown$$
$$C\!=\!C\diagup\!\!\!{}^H$$
$$H\diagup \qquad \diagdown CH\!-\!CH\!-\!CH_2$$
$$\underset{\displaystyle OH}{|}\ \underset{\displaystyle NH_2}{|}\ \underset{\displaystyle OH}{|}$$

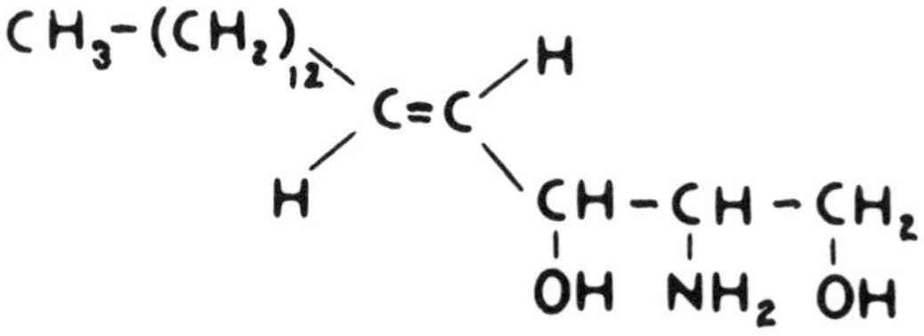

SPHINGOSINE

Dietary Lipids and Postnatal Development
Raven Press, New York © 1973

Human Milk Lipids and Problems Related to Their Replacement

U. Bracco

Department of Research and Development, Nestlé Products Technical Assistance Co. Ltd., Lausanne, Switzerland

I. INTRODUCTION

Many investigations have been carried out in recent years on human milk, applying modern analytical techniques to describe its composition, structure, and absorption pathways. Human milk possesses specific properties such as the ratio of the main chemical constituents (proteins, lipids, carbohydrates) and the composition of the individual group; moreover, its composition is influenced by several factors such as ethnic group, age of the milk (early colostrum, transitional, mature), and dietary carbohydrates and fats. A better knowledge of human milk and the factors influencing its composition is now more and more indispensable in view of its replacement with manufactured infant formulas, which can palliate the increasing deficiency in human milk.

Progress in understanding the chemistry and biochemistry of the main human milk constituents has been described; we limit ourselves here to discussing the problems related to the lipid phase, its composition and structure, and their influence on the intestinal absorption. Cow's milk lipids and other natural fats are also reviewed in order to describe the interrelations between structure and nutritional values and to indicate fat mixtures, which can be introduced as human milk lipids replacements.

II. HUMAN MILK LIPIDS: ISOLATION

Milk lipids exist as microscopic globules emulsified in the aqueous phase of milk. Figure 1 shows some aspects of the fat globule observed with the electron microscope using the freeze-etching technique. The fat globule is about 3μ to 6μ in diameter, and the emulsion is stabilized by an interfacial surface called "fat globule membrane." Absorbed casein is present on the globule, which also contains some lipidic fractions.

23

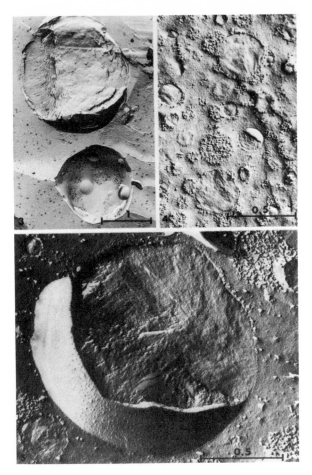

FIG. 1. Electronmicrograph of a milk fat globule.

The investigation of human milk lipids therefore requires a previous separation of the several classes of lipids, namely the free lipids, the fat globule membrane lipids, and the casein-bound lipids. Such separations are obtained in several ways, the most useful of which are reported below.

A. Isolation of Fat Globule and Fat Globule Membrane

As described recently (1), the fat globule and its membrane can be obtained from fresh, "long time-low temperature" (LT-LT process) pasteurized human milk by multistep ultracentrifugation, as indicated in Fig. 2. The sedimented fat globule membrane is collected and freeze-dried: the lipid components of the dry cream and the membrane are extracted with the Folch mixture $CHCl_3$-MeOH (2:1) and freed from nonlipid material by filtration.

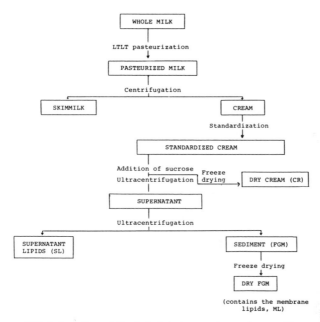

FIG. 2. Isolation of the fat globule membrane from milk.

B. Isolation of Casein-Bound Lipids

Casein is obtained from the skimmed milk by separation of the total complex at pH 4.6 at 20°C and subsequent ultracentrifugation. Freeze-dried casein is extracted first with apolar solvents to eliminate the free fat and subsequently with $CHCl_3$-MeOH (2:1) to obtain the bound lipids.

III. HUMAN MILK LIPIDS: COMPOSITION AND STRUCTURE

The fat globule in human milk is composed mainly of neutral lipids, that is, triglycerides and their intermediary products of biosynthesis or hydrolysis: mono- and diglycerides and free fatty acids.

The fat globule membrane and casein-bound lipids contain large quantities of more polar lipids, that is, phospholipids and sterols; moreover, the fat globule membrane appears to consist of an inner layer of phospholipids and sterols, then a coating of an insoluble protein with an outer covering of a lipoprotein complex, and, finally, adsorbed lipoprotein particles making up an "outer membrane." All these complexes are easily separated by thin-layer chromatography (TLC) on silica gel G, mobile phase petroleum ether-diethylether-formic acid (6:4:0.46). Each fraction is scraped off the plate and transesterified either with dimetoxypropane (2) or a $NaOCH_3/MeOH$ mixture (3); the methylesters obtained are then separated by gas-liquid chromatography (GLC) under standard conditions.

The identification of unusual lipids is carried out either on a coupled system GC/MS or by preparative GC, followed by a mass spectra investigation. The specific distribution of each fatty acid in the glycerides is determined following hydrolysis with pancreatic lipase (4) and subsequent separation and identification of the 2-monoglycerides by TLC-GLC.

Organic phosphorous is determined colorimetrically (5): the separation and identification of each phospholipid is done by TLC and infrared spectrometry. Infrared spectra were recorded using KBr micropellets with a Perkin-Elmer 521 spectrophotometer. Subsequently, each phospholipid is transesterified with sodium methylate as described by Pohl et al. (6), and the resulting fatty acid methylesters are identified by GLC.

Unsaponifiable matter (7) is also characterized by coupled techniques, TLC, GLC, and mass spectrometry.

Table 1 shows the gross composition of human milk lipids according to their position on the fat globule: these are mean values on mature human milk from fresh pooled samples.

It should be noted that partial glycerides and polar lipids are present in larger quantities in the fat globule membrane and in the casein-bound lipids

TABLE 1. *Composition of human milk lipids*

	Fat globule	Fat globule membrane	Casein-bound
	Percent of total lipids		
Triglycerides	98.1	58.2	60–62
Diglycerides	0.7	8.1	2–3
Monoglycerides	tr.	0.6	3–4
Free fatty acids	0.4	7.3	5–5.5
Total fatty acids	91.9	84.5	89.5
Organic phosphorous	0.01	0.9	0.34
Phospholipids (Px26)	0.26	23.4	8.6–8.8
Unsaponifiable	0.31	0.96	1.5

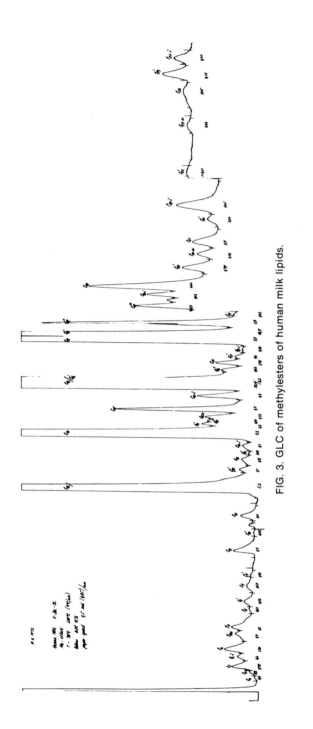

FIG. 3. GLC of methylesters of human milk lipids.

than in the fat globule, which contains a large amount of triglycerides. As far as the fatty acid composition is concerned, more than 142 fatty acids have been identified (8), including even- and odd-numbered, saturated, mono-, di-, and polyunsaturated, and iso-, anteiso-, and multibranched fatty acids.

As an example of the complexity of the fatty acid composition, Fig. 3 shows a chromatogram run with high sensitivity which shows acid components even in small amounts.

Besides the main components, minor constituents have also been identified, including keto- and hydroxyacids, aldehydogenic glycerides, and glyceryl ethers.

Egge et al. (9) have described a large number of branched-chain saturated fatty acids. Kuksis and Breckenridge (10) have given complete tables of fatty acid composition of the triglyceride fractions of human milk fat. By using Scot columns and capillary columns, we were also able to separate some constituents of the $\omega6$ series: structural isomers, *cis-cis, cis-trans, trans-cis,* and *trans-trans* could be evidenced in human milk.

TABLE 2. *Component fatty acids of human milk glycerides*

	Fat globule	Fat globule membrane	Casein-bound
	Percent of total fatty acids		
Saturated			
Below C_{10}	3.1	2.5	1.6
Myristic	8.3	8.6	7.3
Palmitic	24.6	29.2	28.4
Stearic	8.5	9.9	14.2
Arachidic	0.3	0.6	1.8
Above C_{20}	0.2	0.3	2.5
Monounsaturated			
Lauroleic	0.1	0.1	0.1
Palmitoleic	5.9	4.2	0.6
Oleic	33.3	35.6	24.5
Diunsaturated			
Linoleic and isomers	6.7	6.5	9.8
Iso-fatty acids			
Below C_{10}	0.1	0.2	2.2
Me-12Me-tridecanoic	tr.	0.1	0.1
Me-12Me-pentadecanoic	0.1	0.2	0.7
Me-12Me-heptadecanoic	tr.	<0.1	0.6
Me-12Me-nonadecanoic	0.1	tr.	0.4
Tetramethylhexa-decanoic	0.1	<0.1	0.1
Above C_{20}	0.1	0.1	0.1

It should be noted that iso- or anteiso fatty acids are concentrated on the casein-bound lipids rather than on the other lipidic constituents of the milk, and that the high-melting fatty acids are present in larger amounts on the fat globule membrane.

Electronic integration of single peaks gives a complete picture of the fatty acid composition, which we summarize partially in Table 2, giving the amount of the most significant fatty acids.

Triglycerides can also be separated on silver nitrate plates according to their degree of unsaturation; moreover, complete triglyceride spectra can be obtained by direct injection into a gas chromatograph, using short, thermostable packed columns. Table 3 reports the mean values found.

TABLE 3. *Triglyceride distribution and composition in human milk fat globule*

	Percent of total triglycerides		
	SSS	7	
	SUS + USS	36	
	USU + SUU	33	
	UUU	16	
	Tetra-, polyenes	8	

Carbon number	Percent	Carbon number	Percent
18	0.5	40	1.0
20	0.2	42	3.7
26	0.4	44	6.0
28	0.4	46	7.5
30	0.3	48	12.5
32	0.5	50	12.6
36	4.0	52	27.4
38	8.5	54	14.5

Specific distribution of the fatty acids in the triglycerol molecules (fat globule) shows that myristic and palmitic acids, two of the most important saturated acids, are predominantly in position 2, as indicated below:

Fatty acids	Percent of total fatty acids	Percent in position 2
C−12:0	4.6	51
C−14:0	8.3	68
C−16:0	24.6	67
C−18:0	8.5	33
C−18:1	33.3	17
C−18:2	6.7	12

Polar lipids, isolated from the purified and mixed lipids by silicic acid chromatography as proposed by Borgström et al. (11), are separated by two-dimensional TLC; phosphoglycerides and sphyngolipids are the most representative of this class. Their fatty acid distribution shows a high degree of unsaturation: linolenic and arachidonic acids seem to concentrate in the polar moiety of the fat.

Sterol esters, which account for about 15% of the total sterols, also contain a high proportion of octadecadienoic and octadecatrienoic acids. *Cholesterol* and *7-dehydrocholesterol,* odd- and even-numbered, iso-series *hydrocarbons, squalene, tocopherols and tocopherol-like substances, and carotenoids* have also been identified either in the nonsaponifiable fraction or in the molecular distillates. α-Tocopherol, the isomer showing the highest biological activity, has been determined according to the method proposed by Herting and Drury (12).

To give a complete picture of the amount of the nonglyceride substances examined, identified, and determined in human milk lipids, Table 4 gives the values we have found during the last years on several samples.

TABLE 4. *Lipids in human milk*

	Fat globule	Fat globule membrane	Casein-bound
	Percent of total phospholipids		
Phosphatidylethanol		36.6 ± 10	5 ± 10
Phosphatidylcholine		29.7 ± 10	5 ± 10
Phosphatidylinositol		4.6 ± 5	70 ± 10
Sphingomyelin		26.2 ± 10	15 ± 5
Lysolecithin		1.9 ± 15	—
Phosphatidylserine		1 ± 15	5 ± 10
	mg/100 g lipids		
Cholesterol	248	653	710
7-Dehydrocholesterol	6	19	32
Squalene	6	9	8
α-Tocopherol	14	16	n.c.
Carotenoids	0.8	2.2	n.c.

Among the liposoluble components we should mention here vitamin D, which is related to calcium and alkaline phosphatase activity in bone formation, and the antihemorragic vitamin K, which is related to naphthoquinone.

The presence of *lipase* in human milk was confirmed by a rapid increase in free fatty acids; casein-bound lipids showed a rapid spontaneous lipolysis, splitting about 30% of the total fatty acids in a short time.

The liberation of free fatty acids induced a rapid drop in the pH of the milk, which was not observed in other milks, such as cow's milk; the buffering capacity of human milk probably differs markedly from that of cow's milk, as described by Tarassuk et al. (13).

IV. CHANGES IN HUMAN MILK LIPIDS RELATED TO SEVERAL FACTORS

The mean values shown here and corresponding to a mature milk are subject to wide variations depending on physiological or psychological individuality, ethnic group, age of the milk, dietary carbohydrates and fats, and environmental conditions. With regard to the influence of the age of the milk, Fig. 4 shows the variation of total fat, some fat-soluble vitamins, and fatty acid composition observed on colostral (1 to 5 days old), transitional (6 to 10 days old), and mature milks (more than 30 days old). In comparison with mature milk, colostral milk shows significant differences in the lipid fraction, namely, a lower amount of total lipids, higher concentrations of vitamin A and carotenoids, a higher amount of vitamin E and tocopherol-

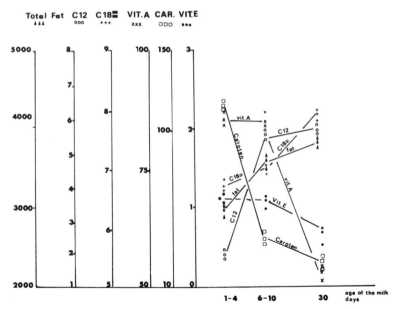

FIG. 4. Human milk lipids: change in composition related to the age of milk. Average values: fat, vitamin A, and carotenoids as mg/100 ml milk; vitamin E as μg/100 ml milk; C_{12} and C_{18} as g/100 g fat.

like substances, a lower content of short-chain fatty acids and lauric acid, and a lower amount of octadecadienoic acid and a higher content on arachidonic acid. An increase in total lipid phosphorous was also observed during the lactation.

Colostrum particularities have special significance for the newborn: low fat and high concentrations of fat-soluble vitamins are strictly correlated with the presence of enzymes in the newborn and with its physiological needs.

Diet can also influence human milk lipids: Insull (14) demonstrated a significant increase in octadecadienoic acid correlated with a high amount of linoleic acid in the mother's diet (corn oil). Read et al. (15) found a significant influence of dietary carbohydrates on the fatty acid composition of the milk, in that they related a high dietary intake of carbohydrates to high lauric and myristic acid levels in the milk and reduced quantities of C_{18} acids. Moreover, the caloric values of the diet seem to influence the lipid content of the milk: an increase in fats has been observed for diets enriched from 2000 to 3000 cal/day.

V. HUMAN MILK LIPIDS REPLACEMENT

The manufacturing of infant feedings requires not only a wide knowledge of human milk lipids and their nutritional significance, but also a systematic investigation of the characteristics of other natural lipids, either from animal or from vegetable sources.

A. *Cow's milk* is largely used in human nutrition; its utilization is related to its composition. High proportions of several nutrients, including Ca and P ions, vitamin D, B-complex vitamins, proteins, riboflavin, and ascorbic acid, are contributed by cow's milk to the recommended daily dietary allowances.

Chemical characteristics (16), (17), physical structure (18), and biogenesis (19) have been widely investigated in cow's milk lipids. Moreover, constituent fatty acids of milk fats of additional species have recently been described (20). Without listing all the characteristics of cow's milk lipids, we will point out the most significant differences we and other authors have found between human and cow's milk fat. These differences may have important nutritional significance with regard to lipid digestibility.

In comparison with human milk lipids, cow's milk lipids show: a higher content of short-chain (C < 10) fatty acids, namely butyric acid; a higher proportion of total saturated acids and methyl-branched saturated acids (i.e., phytanic acid); the presence of monoethenoic acids of *trans* configuration, the best known being "vaccenic acid" (*trans*-11-octadecenoic acid); the presence of monoethenoic acids in which the double bond occupies other positions than the 9, 10 position [i.e., the dec-9-enoic acid CH_2=CH-$(CH_2)_7 COOH$]; a lesser amount of diethenoic acids (i.e., linoleic acid) and their structural and geometrical isomers; a less preferential distribution of palmitic acid in the 2 position of the triglyceride molecule; a lower level of total organic phosphorous, namely phosphatidylethanolamine and phosphatidylinositol; a lower degree of unsaturation in the phospholipid moiety; and smaller amounts of cholesterol and 7-dehydrocholesterol.

Fat-soluble vitamins are also contained in different proportions in both milks: carotene and vitamin A are present in smaller amounts (40 μg and 30 μg/100 ml milk, respectively) in cow's milk; the α-tocopherol mean values are also largely lower in cow's milk lipids. The concentration of this vitamin is related to that in the diet and it is subjected to important seasonal variations; nevertheless, in mature cow's milk it is always not more than one-fifth the mean content of human milk fat.

Herting and Drury (12) also give the ratio α-tocopherol/polyunsaturated fatty acids (mg/g); we have calculated average values of 1.6 and 0.6, respectively, for human and homogenized cow's milk lipids.

Cow's milk fat is also influenced by the diet to a lesser extent than is human milk; in fact, rumen bacteria act either as hydrogenation factors or as sources of volatile iso-fatty acids (i.e., iso-butyric acid, iso-valeric acid, etc.) modifying the lipid composition of the diet.

Recent work to increase the polyunsaturated fatty acids of cow's milk fat has been done by Scott et al. (21) and Plowman et al. (22), by feeding cows a diet containing a formaldehyde-treated safflower oil-casein particle, to protect the linoleic acid from the bacterial activity in the rumen. Cow's milk fats containing up to 30% linoleic acid are reported.

B. *Vegetable fats or oils* have been introduced, for a number of years, as cow's milk fat replacers: they are claimed to be carriers of natural polyunsaturated fatty acids and therefore seem to be ideal modifiers of cow's milk lipids for dietetic purposes.

Apart from infant foods, products are now on the market that are classified under the names "filled milks" and "imitation milks." The former are based on a combination of oils other than butter oil with milk solid nonfats; the latter also contain vegetable fats mixed with natural nondairy solids.

Vegetable oils, with the exception of coconut and palm kernel oil, show high amounts of mono-, di-, and trienes and consequently the triglycerides are mainly of mono-, di-, and tripolyunsaturated type.

Generally, only a small percentage of the total palmitic acid is present in the 2 position of the triglyceride molecule: saturated fatty acids having a chain length of greater than 18 carbons are found predominantly in the 1 and 3 positions. Excepting soybean oil, the amount of total phospholipids is also relatively low in comparison with that of human and cow's milk fat.

VI. NUTRITIONAL VALUES CORRELATED WITH THE STRUCTURE OF NATURAL FATS

The characteristics of the lipid fraction in infant nutrition, that is, composition, minor constituents, and structure, are strictly correlated with the nutritional value and the physiological significance of a diet.

Lipids play an important role not only with regard to the digestion and absorption of a food, in terms of energy suppliers, but also in relation to several factors involving the nature of serum and depot lipids, the role of essential fatty acids, and the degree of absorption of nutrients such as calcium ions and fat-soluble vitamins.

It is generally agreed that human milk lipids meet the nutritional requirements of the newborn (23), (24) and therefore that each infant formula should show characteristics similar to those of human milk fat in its fat composition.

The superior absorption of the fat of human milk over that of cow's milk can be explained by several factors. First, the smaller amount of trisaturated glycerides allows a better absorbability of human milk; the relationship between fat absorbed and its melting point has been largely investigated by Deuel (25) and Mattson (26). In a fat balance study it was demonstrated that the absorbability decreases as the amount of trisaturated glycerides increases. Second, the glyceride structure plays an important role in fat absorption. The specific positional distribution of long-chain saturated fatty acids, namely palmitic acid, may confer better digestibility to human milk fat.

Freeman et al. (27) and Fomon (28) related the digestibility of fats used in infant feedings to the specific distribution of palmitic acid in the 2 position of the triglyceride molecule. Extensive work in this respect has been done by Tomarelli et al. (29), who examined the relationship between fat absorption and positional distribution of fatty acids. They conclude that the unsaturated and short-chain saturated fatty acids are well absorbed regardless of structural distribution, whereas the palmitic acid absorption is well correlated with the position of the fatty acid in the triglyceride. Particularly high amounts of palmitic acid in the 2 position account for a better absorbability of a fat. Lard, for example, which contains a large amount of palmitic acid ($\sim$ 85% of the total) in the 2 position, possesses a high degree of absorption, whereas randomized lard, in which palmitic and stearic acids are equally distributed on the glyceride molecule, shows a significant loss in absorption (30).

Sweeney et al. (31) and Takizawa (32) studied the differences in fat depot composition in relation to diet. Recently, Ballabriga et al. (33), by using diets containing different amounts of *cis,cis*-dienoic acid, showed a close relationship between the nature of fat and the proportion of *cis,cis*-dienoic acid in blood and in depot fat. Their results show that milk formulas containing 20% vegetable fat in the form of corn oil produce values of linoleic acid in blood and depot fat very close to the values obtained with human milk fat.

Furthermore, dietary lipids influence the composition of the nonsaponifiables in the serum lipids, for example, the amount of cholesterol (34).

Pickering et al. (35) studied the influence of fatty acid composition of infant formulas on the development of arteriosclerosis and on the lipid composition of blood and tissues in monkeys. They compared human and cow's milk fat: although cow's milk fat gives higher serum cholesterol, lipid phosphorous, and "globulin-lipoprotein fraction" levels, there was essentially no difference between the vascular lesions of the two groups.

Greenberg and Wheeler (36) studied cow's milk fat and human milk fat and compared them with a fat formula containing 10% linoleic acid. The linoleic acid level in plasma, erythrocytes, and tissues was related to the amount of the acid in the diet: the hypercholesterolemic effect of cow's milk fat was confirmed.

VII. FORMULATION OF LIPID PHASES FOR INFANT FORMULAS

It will be apparent from the above discussion that in the formulation of human milk fat-like lipid phases several factors must be taken into consideration, such as fatty acid composition, palmitic acid distribution, triglyceride structure, sterol, and phospholipid and liposoluble vitamin levels. This represents only the gross composition of a fat: nonglyceride substances, alkyl ethers, alkyl-enyl ethers, and related plasmalogens and long-chain bases described as minor components of human milk fat probably possess an important physiological significance which is at present largely unknown.

We have an abundance of information about human milk fat, but we also lack exact knowledge of certain points. At the present stage of our knowledge and to formulate suitable fat mixtures, we could consider the ranges of values reported in Table 5. Fat blends showing some or all of the characteristics discussed above can be prepared in several ways. We could use a simple blend of butter oil and vegetable fats with an increased poly-

TABLE 5. *Minima-maxima values for human milk lipid-like fat mixtures*

Main fatty acid composition	Percent of total fatty acids
C4:0	<3
C6:0	<1
C8:0	<1
C10:0	<2
C12:0	2–6
C14:0	6–10
C16:0 (max. in β position)	24–30
C16:1	2–3
C18:1	32–38
C18:2	2–12
C18:3	<2
C20:4	<2
Triglyceride composition	Percent of total triglycerides
SSS	5–10
SSU	30–40
SUS	5–10
SUU	20–30
USU	10–30
UUU	5–10
Cholesterol, mg/100 g fat	<600
Phospholipids (calc. as lecithin) mg/100 g fat	<1400

unsaturate level, or more sophisticated mixtures such as those examined by Tomarelli et al. (29), which contain oleo oil, corn oil, coconut oil, soybean oil, or lard. Finally, we could computerize optimal solutions starting from the characteristics of several vegetable and animal fats and from the expected characteristics of the fat blend. One of the computer-optimized solutions we obtained was a fat mixture containing liquid oil (i.e., olive and soybean oil), vegetable fats (coconut, palm kernel, and palm oil) and animal fats (lard and cow's milk fat).

On the other hand, a draft standard for infant formulas was submitted to the Committee of the Codex Alimentarius for Dietetic Products (37). This standard applies to food in liquid or powdered form intended for use as a substitute for human milk in meeting normal nutritional requirements. In the case of lipids, it states that the product shall contain linoleate (in the form of triglycerides containing linoleic acid) at a level not less than 300 mg, expressed as linoleic acid per 100 available calories, and the following values for the fat-soluble vitamins:

Vitamin A   minimum 250 IU
or 75 μg expressed as
retinol, or 450 μg
expressed as β-carotene

Vitamin D	minimum 40 IU; maximum 40 IU
Vitamin E	minimum 0.3 IU or
	0.3 mg expressed as
	α-tocopherol

All data was calculated per 100 available calories. The limits of total lipid content were set to be between 2 g and 6 g per 100 available calories.

The formulation of new infant formulas requires not only improved knowledge of the lipid field, but also considerable research on the dietary requirements of infants, on digestion and absorption of lipids, and on their metabolism and transport. For many years we have carried out a collaborative study in this field with pediatricians, chemists, and experts on electron microscopy. Problems to be solved are the micellar solubilization either at the oil/water interface or at the level of the intestinal cell; the preferential solubilization of monoglycerides; and the mechanism of resynthesis of lipids in the intestinal mucosa.

Lipid digestion *in vivo* can be followed at present by electron microscopy either by negative staining, freeze etching, or electron diffraction. Some micellar structures are of the lamellar phase type: during the course of the digestion the thickness of this distribution diminishes (i.e., from 40 Å to 18 Å), giving typical structures of mono-, di-, and triglycerides and of free fatty acids. Moreover, electron diffraction is now applied in the lipid field, providing a tool for the characterization of microgram quantities of crystalline material.

This collaborative study is the best way to clarify the triglyceride pathway of absorption and to judge the effectiveness of several fat compositions in infant feeding.

VIII. CONCLUSIONS

In spite of significant variations, human milk lipids possess well-defined physical and chemical characteristics as far as the main components are concerned. On the other hand, the structures and physiological significance of minor lipid constituents are at present largely unknown. In assessing new fat phases for infant formulas we should ensure that they meet the dietary requirements of the newborn in terms of energy, intestinal absorption, composition of body lipids, and influence on the absorption of other nutrients. Structure and composition of human milk fat-like lipids can be calculated on the basis of present knowledge, and their metabolism can be followed by new analytical techniques. Nevertheless, only with practical experimentation and clinical trials can a final judgment on the physiological value of new human milk replacers be given.

REFERENCES

1. Bracco, U., Hidalgo, J., and Bohren, H. (1972): *J. Dairy Sci.*, 55:165.
2. Mason, M. E., and Waller, G. R. (1964): *Anal. Chem.*, 36:583.
3. Shehata, A. Y., de Man, J. M., and Alexander, J. C. (1970): *Can. Inst. Food Technol. J.*, 3:85.
4. Luddy, F. E., Barford, R. A., Herb, S. F., Magidman, P., and Riemenschneider, R. W. (1964): *J. Am. Oil Chem. Soc.*, 41:693.
5. International Union of Pure and Applied Chemistry (1966): *Fats and Oils. Method II.D.16: Determination of Phosphorous in Oils and Fats: Standard Methods for Analysis.* Butterworths, London.
6. Pohl, P., Gasl, H., and Wagner, H. (1970): *J. Chromatogr.*, 40:425.
7. International Union of Pure and Applied Chemistry (1964): *Determination of the Unsaponifiable Matter. Method II D. 5.*, London.
8. Jensen, R. G., Quinn, J. G., Carpenter, D. L., and Sampugna, J. (1967): *J. Dairy Sci.*, 50:119.
9. Egge, H., Murawsky, U., and Illiken, F. (1968): *Hoppe-Zeyler's Z. Phys. Chem.*, 349:4.
10. Kuksis, A., and Breckenridge, W. C. (1968): In: *Diary Lipids and Lipid Metabolism*, edited by M. F. Brink and D. Kritchevsky. Avi Publishing Company, Westport, Conn.
11. Borgström, S., Borgström, B., and Rottenberg, M. (1952): *Acta Physiol. Scand.*, 25:120.
12. Herting, D. C., and Drury, E. J. (1969): *Am. J. Clin. Nutr.*, 22:147.
13. Tarassuk, N. P., Nickerson T. A., and Yaguchi, M. (1964): *Nature*, 201:298.
14. Insull, W. (1959): *J. Clin. Invest.*, 38:443.
15. Read, W. W. C., Lutz, P. G., and Tashjian, A. (1965): *Am. J. Clin. Nutr.*, 17:180.
16. Morrison, W. R. (1970): In: *Topics in Lipid Chemistry*, edited by F. D. Gunstone. Logos Press, London.
17. Privett, O. S., Nutter, L. J., and Gross, R. A. (1968): In: *Dairy Lipids and Lipid Metabolism*, edited by M. F. Brink and D. Kritchevsky. Avi Publishing Company, Westport, Conn.
18. Kurtz, F. E. (1965): In: *Fundamentals of Dairy Chemistry*, edited by B. H. Webb and A. H. Johnson. Avi Publishing Company, Westport, Conn.
19. Storry, J. E., and Rook, J. A. F. (1964): *Biochem. J.*, 91:27C.
20. Glass, R. L., and Jenness, R. (1971): *Comp. Biochem. Physiol.*, 388:353.
21. Scott, T. W., Cook, L. J., Ferguson, K. A., McDonald, I. W., Buchanan, R. A., and Loftus Hills, G. (1970): *Austr. J. Sci.*, 32:291.
22. Plowman, R. D., Bitman, J., Gordon, C. H., Dryden, L. R., Goering, H. K., and Wrenn, T. R. (1972): *J. Dairy Sci.*, 55:204.
23. National Research Council (1953): *Bull. Nat. Res. Coun.*, No. 254, Washington, D.C.
24. Macy, I. G., and Kelly, H. J. (1961): In: *Milk: The Mammary Gland and Its Secretion*, edited by S. K. Kon and A. T. Cowie. Academic Press, New York and London.
25. Deuel, H. J. (1955): In: *The Lipids*. Interscience, New York.
26. Mattson, F. H. (1967): In: *Proceedings of the 1967 Deuel Conference on Lipids, on the Fate of Dietary Lipids*, edited by G. Gorngill and L. W. Kinsell. U.S. Dept. of Health, Education and Welfare, Washington, D.C.
27. Freeman, C. P., Jack, E. L., and Smith, L. M. (1965): *J. Dairy Sci.*, 48:853.
28. Fomon, S. J. (1967): In: *Infant Nutrition*. Saunders, Philadelphia.
29. Tomarelli, R. M., Meyer, B. J., Waeber, J. R., and Bernhart, F. W. (1969): *J. Nutr.*, 95:583.
30. Renner, R., and Hill, F. W. (1961): *J. Nutr.*, 74:254.
31. Sweeney, M. J., Etteldorf, J. N., Throop, L. J., Timma, D. L., and Wreen, E. L. (1963): *J. Clin. Invest.*, 42:1.
32. Takizawa, Y. (1967): *Acta Paediatr. Jap.*, 9:87.
33. Ballabriga, A., Martinez, A., and Gallart-Catala, A. (1972): *Helv. Paediatr. Acta*, 27:91.
34. Campbell, R. G., Hashim, S. A., and von Itallie, T. B. (1968): In: *Dairy Lipids and Lipid*

Metabolism, edited by M. F. Brink and D. Kritchevsky. Avi Publishing Company, Westport, Conn.

35. Pickering, D. E., Fisher, D. A., Perley, A., Basinger, G. M., and Moon, H. D. (1961): *Am. J. Dis. Children,* 102:72.
36. Greenberger, L. D., and Wheeler, P. (1968): In: *Dairy Lipids and Lipid Metabolism* edited by M. F. Brink and D. Kritchevsky. Avi Publishing Company, Westport, Conn.
37. Codex Alimentarius for Dietetic Products (1971): *Step 8 of the Procedure for the Elaboration of Worldwide Standards.* Proposed Draft Standard, Ali. 71/26- Appendix IV.

Dietary Lipids and Postnatal Development
Raven Press, New York © 1973

Structural Lipids and Their Polyenoic Constituents in Human Milk

M. A. Crawford, A. J. Sinclair, *P. M. Msuya, and *A. Munhambo

*Nuffield Institute of Comparative Medicine, Zoological Society of London, Regent's Park, London, N.W.1., England and *Department of Biochemistry, Faculty of Medicine, University of Dar-es-Salaam, Dar-es-Salaam, Tanzania*

I. INTRODUCTION

Most texts present nutrition with the focus on protein; foods are either rich in protein or rich in calories. The state of food and agriculture is described in terms of protein availability. Malnutrition in Africa is oversimplified to "protein" malnutrition. Nutritional advice centers on the protein-rich foods, and rapidity of body growth in the laboratory animal is equated with a notion of excellence.

This type of thinking has led to the widely held view that the higher the protein content of a nation's food supply, the bigger, the better, and the healthier that nation will be. Similar concepts are applied to the feeding of children and, as a result, protein has been endowed with an almost mystic property. It so happens that natural foods rich in protein are also rich in other nutrients which are overlooked in these repeated oversimplifications.

Some consider the human species to have reached a degree of evolutionary excellence measurably superior to that of other species, yet if we look at the composition of milks we find that there is less protein in human milk than in that of practically any other species. The rat has 9 g/100 ml, the dog 8 g, the hippopotamus 7 g, the pig 6 g, the ox 4 g, and man 1 g (1). During suckling the mean growth rate of the pig is 295 g/day, of the calf 580 g/day, and of the human infant only 25 g/day.

Such figures do not imply that the human species is an example of outstanding quantitative protein nourishment!

The remarkably small proportion of protein present in human milk stands as an unexplained contradiction to current nutritional thinking. However, current concepts have yet to embrace the fact that the fundamental distinction between species involves important qualitative elements; these may in turn arise from a blend of individual quantitative parameters. The indi-

41

vidual components of the skeletal and muscular architecture in the large land mammals is sufficiently alike for a common anatomical vocabulary to be employed throughout. The femur is common, although there are variations in size and shape relative to the rest of the body. The differences are qualitative.

In the qualitative sense, the human species is usually thought of as superior to other mammals, yet there are many biological attributes of other species which could well be considered superior to those of man; for example, the development of musculature in the antelopes and bovids, the night vision of the cats, and the remarkable agility of the tree-living primates.

The feature which is outstanding in man is unquestionably the brain and the peripheral nervous system. It is clear that whereas protein is quantitatively the most important structural factor in muscle, lipid is the most important in the brain and peripheral nervous system.

Bearing the chemistry of the brain in mind, the apparent paradox of 6% protein in pig's milk, 4% in cow's milk, and only 1% in human milk can be resolved if one discards the equation between maximum growth and excellence and substitutes brain growth, or what Widdowson (1) describes as the "harmony of growth." Since the brain is predominantly a lipid-rich structure, a harmony of growth implies at least a proper balance between lipid and protein nutrients and indeed, for man, the lipids may be just as important as protein.

Such an alteration from the conventional view of nutrition introduces elements of interrelated nutrients rather than isolated "protein," and qualitative factors rather than quantitative measures of body growth that omit brain growth.

Measurements of brain growth using DNA have shown that most of the cells have accumulated in the human brain by birth, leaving about 20 to 30% of the total cell division to be completed in the first postnatal 6 months (2). The significant increase in brain weight which occurs subsequently is due to some extent to an increase in the cellular content of RNA, protein, and lipid, but mainly to the development of myelin, although extrapyramidal cells in the brain and nervous system must continue to appear until very much later. The gray matter, which is mainly cellular material, is about 50%, and the white matter, which is mainly myelin, is about 70% structural lipid on a dry-weight basis. Consequently, lipid factors will be important to postnatal development.

Since much of the brain cellular lipid is "essential," the consideration of milk in the light of the dynamic biochemistry of a developing neonate could be of special importance to communities that have embarked on a course of substituting cow's milk or commercially manufactured products for human breast. Attempts have been made, notably by SMA, to persuade an uninformed public that a collection of proteins, minerals, vitamins, and some vegetable oil is as nutritious as human milk. With such products, problems have been documented in the past both with regard to mineral absorption and lipid absorption (3–5) and from severe hemolytic anemia resulting from an incorrect balance of tocopherol and polyenoic content (6–8). If a mineral imbalance during bone development can affect bone structure, the possibility must exist that an incorrect lipid balance could influence the brain during organogenesis.

The crude relevance of nutrition during organogenesis has already been observed. Paoletti and Galli (9) have shown that maternal lipid malnutrition in rats results in an irreversible restriction of learning ability; rats fed on a diet low in lipid show no external symptoms of essential fatty acid (EFA) deficiency, but have radically altered liver lipids, reduced brain size and cell number, and low survival rate (12). In the human, malnutrition in early childhood has been described by Cravioto et al. (10) as leading to a permanent impairment of intellectual development. We have shown that the phase of cell division in the rat brain is associated with an accumulation of arachidonic (C20:4, n - 6) and docosahexaenoic (C22:6, n - 3) acids (11). These are long-chain polyenoic acids (LCP) derived from the parent short-chain polyenoates (SCP) linoleic (C18:2, n - 6) and linolenic (C18:3, n - 3) acids, respectively.

The LCP can be obtained either preformed in animal foods including milk (13) or by biosynthesis from the SCP. This does not mean that the SCP are necessarily equivalent biologically to the LCP, since it has been shown that LCP are incorporated more effectively into rat liver lipids (14).

Most analyses of the fatty acids in human milk describe a triglyceride pattern which does not extent beyond linoleic acid (15), but in confirmation of an earlier report by Insull and Ahrens (16) we have found that the triglycerides of human milk do contain both SCP and LCP and feel that in view of the significance of the polyenoic acids in the brain we should examine the occurrence of these polyenoic constituents in human milk.

II. METHODS

Our methods for fatty acid analyses have been described previously (12) and basically employ gas-liquid chromatography (GLC) after the separation of the lipid classes on thin-layer chromatography (TLC) for triglycerides and countercurrent separation for phosphoglycerides prior to TLC for separation of the individual classes.

The human studies were made with the cooperation of wives of members of the staff at our Institute, the Salvation Army Mothers' Hospital, Clacton, London, and Muhimbili Hospital at the Medical School, Dar es Salaam.

III. CRUDE MILK COMPOSITION

The cream can be separated from the milk serum by centrifugation, and it is of interest that the cream is predominantly triglyceride (TG) whereas the phosphoglycerides are concentrated in the serum, probably as lipoprotein (Table 1). The phosphoglyceride pattern is similar to that of cellular phospholipids, hence our reference to structural lipids: there were approximately equal proportions of ethanolamine (EPG), choline phosphoglyceride (CPG), and sphingomyelin (SPH). This pattern is unlike human plasma, which is predominantly CPG. Table 1 includes data from the literature and is essentially in agreement with data reported by Morrison (17, 18).

TABLE 1. *Composition of human milk*

CRUDE			
Total Solids	Protein	Fat	Phospholipid
12.4–13.0 g/100 g	1.1–1.7 g/100 g	4.2–4.7 g/100 g	0.16–0.2 g/100 g

FAT
Percent of total fat in

	Serum	Cream
Triglyceride	45–55	80–85
Diglyceride and monoglyceride	2.2–6	2.6–8
Free fatty acids	4–10	3–5
Phospholipid	33–51	4–6
Protein	1.2–1.8 g/100 ml	0.2–0.3%

PHOSPHOLIPID
$n = 6$ (7–21 days)

	Mean $\pm$ S.E. (%)
Ethanolamine phosphoglyceride	28 $\pm$ 3.4
Choline phosphoglyceride	30 $\pm$ 3.8
Serine phosphoglyceride	5.3 $\pm$ 0.2
Inositol phosphoglyceride	4.0
Sphingomyelin	32 $\pm$ 4.0

IV. POLYENOIC CONTENT OF MILK FRACTIONS

The mean linoleic acid in 32 triglyceride samples was 8.5%. The EPG and CPG contained a higher proportion of linoleic, and the major LCP derivative from linoleic was arachidonic. The principal linolenic LCP was docosahexaenoic acid (Fig. 1), which is not the case in all mammals, as we shall see later.

We have calculated the relative proportions of the two individual fatty acid families (*n* - 6 and *n* - 3). The *n* - 3 family is often neglected because linolenic acid is frequently present in small quantities in comparison with linoleic acid, and the C20 and C22 derivatives can easily be missed because of their long retention times on GLC. The main LCP *n* - 3 acid in human milk is docosahexaenoic, and the sum of all LCP *n* - 3 acids is usually quite significant even if the individual contribution of docosapentaenoic or docosahexaenoic acids may at first sight seem small.

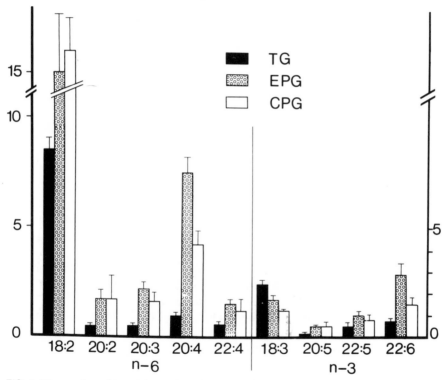

FIG. 1. Mean polyenoic content of human milk ($n = 32$). The triglyceride content of human milk is so large in relation to the phosphoglyceride content that for practical purposes the total fatty acid composition is close to that of the triglyceride fraction. Although small proportions of long-chain derivatives of linoleic and linolenic acids of the order of 0.3 to 1.5% may individually seem small, the sum total of these constituents represents a significant proportion of the total calories. In this preliminary analysis it is likely that C18:3, C20:4, and C22:4 of the TG's are contaminated with C20:1, C22:1, and C24:1, respectively. A detailed report is in preparation.

In Table 2 it can be seen that the EPG fraction is the richest with respect to polyenoic acids, which amount to 34% of the total fatty acids and aldehydes compared to 29% for the CPG and 14.8% for the triglycerides. The total $n - 6$ and $n - 3$ LCP acids was highest in the EPG fraction at 17%.

TABLE 2. *Summary of human milk polyenoates*

		Total		LCP		Ratio
		$n - 6$	$n - 3$	$n - 6$	$n - 3$	$n - 6/n - 3$
Triglyceride	Mean	11.0	3.9	2.4	1.5	2.8
	S.E.	0.9	1.5	0.4	0.3	
Ethanolamine phosphoglyceride	Mean	28.0	6.3	13.0	4.5	4.4
	S.E.	4.3	0.8	1.6	0.8	
Choline phosphoglyceride	Mean	25.0	4.1	8.7	2.9	6.1
	S.E.	4.0	1.0	2.6	0.8	

The ratio of $n - 6$ to $n - 3$ acids varied, being 6 to 1 in the CPG fraction and 4.4 and 2.8 to 1 in the EPG and TG.

V. VARIATION IN MILK EPG, CPG, AND TG

The most remarkable feature of the polyenoic content of milk was its variation. In Fig. 2 we have plotted the data from two samples of milk at 3 weeks after delivery, one of which was rich and the other poor in polyenoic acids. The least variation was in the linoleic acid. Marked differences occurred in linolenic and also in the $n - 6$ and $n - 3$ LCP derivatives of EPG and CPG. Of special interest was the high proportions of C20:2, $n - 6$ and C20:3, $n - 6$ along with a low proportion of arachidonic acid (C20:4, $n - 6$) in both the EPG and CPG fractions, together with a low proportion of docosahexaenoic acid in milk which had a small total proportion of polyenoic acids.

VI. COMPARISONS OF MILK TRIGLYCERIDES FROM EUROPEAN AND AFRICAN SUBJECTS

We made a comparison of the polyenoic fatty acids in 15 samples of milk triglycerides from Europeans and nine samples from Africans taken between

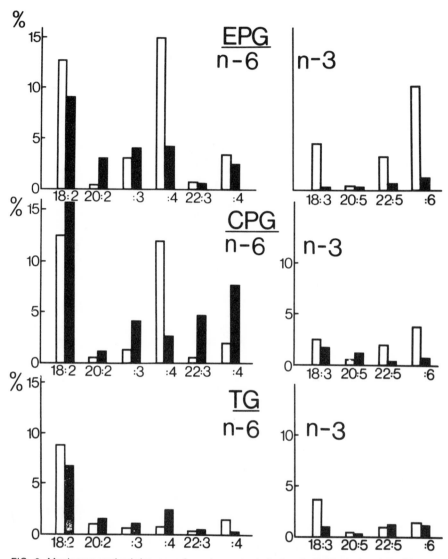

FIG. 2. Maximum and minimum polyenoic contents in two individual samples of human milk.

14 and 20 days of lactation, and so far the only significant difference to emerge (Fig. 3) has been a higher proportion of linoleic acid in the Africans (10.2 ± 1.1 versus 7.6 ± 0.36). Some of the early colostrum triglycerides from the Africans contained as much as 15% linoleic acid.

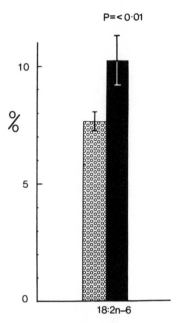

FIG. 3. Significant differences were observed in the linoleate content of milk triglycerides in European (first column) and Tanzanian (second column) mothers. The samples were matched as far as possible for time after parturition.

Studies are in progress to extend these comparisons to the phosphoglyceride fractions.

VII. SPHINGOMYELIN

The data on the fatty acids of sphingomyelin from 21 milks at 14 to 40 days of lactation are presented in Fig. 4. The mean data gives a false impression of an almost even distribution of the C20, C22, and C24 saturated and monounsaturated fatty acids. Individual samples would either be rich in the C20 or the C22 or the C24 acids, which when averaged gives an impression of uniformity. In established lactations of 3 to 5 weeks duration, the lignoceric (C24:0) and nervonic acids (C24:1) were the major long-chain acids, and in these samples the C20 and C22 monounsaturated acids appeared in relatively small proportions. The C20 and C22 acids were frequently raised and the C24 acids lowered in early milks except the colostrum (see Section I). In all samples studied, the nervonic (C24:1) acid was greater than lignoceric acid (C24:0), whereas the saturated C20 and C22 acids were greater than the corresponding monounsaturated acids.

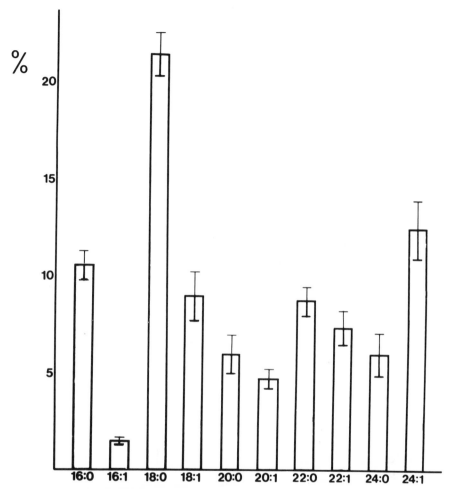

FIG. 4. Sphingomyelin fatty acids of human milk ($n = 21$). The text describes how average figures for the individual fatty acids of milk sphingomyelin are not representative of the pattern found in individual samples, which tend to be rich either in C24 and C24:1 or, alternatively, in the C20 and C22 acids.

VIII. SEQUENTIAL ANALYSIS OF MILK EPG

We had the opportunity of following the composition of milk from the colostrum through to 2 months. We analyzed six cases, all of which gave different results, but underlining these differences a common principle appeared throughout.

The colostrum was always richer in its polyenoic content, of all fractions, than subsequent samples.

In four of the six samples there was a marked loss of polyenoic content in the 7-day sample and in two, this polyenoic depression was extended to the sample on the 14th day. In particular, the C20:2, n - 6 and C20:3, n - 6 rose and the C20:4, n - 6 fell, but by the second and third weeks the more normal balance of a high C20:4, n - 6 and low proportions of intermediates was restored.

In Fig. 5 we have plotted the changes in EPG polyenoics in one case in which this apparent polyenoic deficiency was observed in both the 7- and 14-day samples and subsequently was restored. The changes seen are not unlike those seen in polyenoic deficiency, but we did not detect any significant amount of n - 9 acids in these cases.

This accumulation of intermediates and loss of arachidonic and docosa-

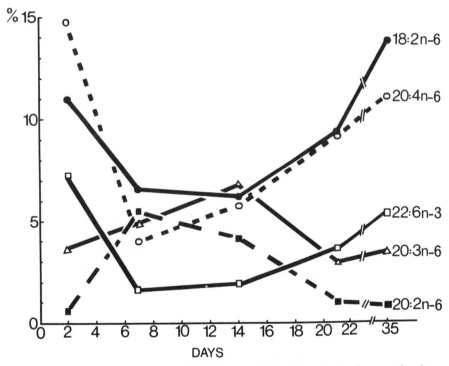

FIG. 5. The sequential change in milk EPG composition from one mother, starting from the colostrum. The postcolostrum fall in principal polyenoics was accompanied by a rise in intermediates; little evidence of C20:3 (n - 9) has so far been detected during this period.

hexaenoic in the milk immediately after the colostrum might well be a deficiency syndrome resulting from the very substantial requirements of LCP to endow the developing fetal brain and liver with its appropriate requirements.

IX. SEQUENTIAL ANALYSIS OF SPH

In Fig. 6 it can be seen that the C24:0 and C24:1 in sphingomyelin was considerably lower in the 7 to 14 day samples than in the colostrum and at 35 days. At the same time there was a parallel rise and fall in the C18 and C22 acids.

In general, the LCP acids in the phosphoglycerides from the cases we

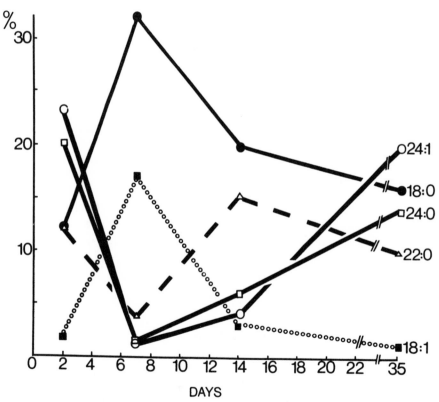

FIG. 6. The sequential change starting with the sphingomyelin fatty acids of colostrum from the same mother as in Fig. 5. It is interesting that the C24 and C24:1 were restored to their colostrum levels despite a marked loss in the first week.

studied did not return to their high colostrum levels, but the C24:0 and C24:1 of SPH did recover. Although there are obviously sweeping hormonal changes in progress in the mother during this postnatal period, and these may in part be responsible for the alterations in chain-elongation products, it is also likely that nutritional factors during pregnancy may be especially important. No guidance is given regarding foods which could supply structural lipids. In this respect it was of interest that one mother who had been under strict dietary control involving calorie restriction to avoid weight gain during pregnancy exhibited the same postnatal loss of chain-elongation products, followed by a loss of phosphoglyceride and then cessation of lactation in the fourth week, despite her own desire to continue to feed her baby and a previously successful lactation following a pregnancy in which dietary restriction had not been imposed.

X. POSSIBLE IMPLICATIONS TO NEONATE NUTRITION

In experiments with rats it is clear that diet does affect milk composition. The impact of diet on maternal milk lipids can be partially understood but little is known about neonate nutrition, and it is worth examining the possible implications to the neonate of the polyenoic acids in human milk.

It is generally assumed that linoleic acid will satisfy all the lipid nutrient requirements for man, but the foundation for this assumption is based on experiments in rats, which have almost invariably used vegetable oils that also contain linolenic acid in small amounts. Furthermore, it is said that babies reared on artificial milks grow fat and do well. This, of course, is an entirely subjective comment, and in view of Osborn's evidence (19), small as it is, that infants reared on artificial foods were found to have pronounced early atherosclerotic changes, one must ask if apparent external fitness need be an indication of internal fitness in the vasculature, nervous system, and developing organs.

We would like to question the validity of the notion that linoleic is the only essential nutrient on three considerations:

1. *Metabolic:* It has been known for some time that docosahexaenoic acid is incorporated more efficiently into liver phosphoglycerides than its precursor linolenic (20). If this is the case, the LCP will be biologically different to the SCP, and the question must be asked whether certain advanced mammalian species have developed a requirement for the LCP. Such a requirement need not be absolute in the sense of a total absence of LCP resulting in failure to maintain life, because many species exist without a dietary source of LCP, but bearing in mind the important qualitative evi-

dence in the distinction between species, a qualitative function for LCP in a complex species such as man becomes entirely plausible.

2. *Analytical:* A ratio between 3 and 4 to 1 has been reported for the $n - 6$ and $n - 3$ acids in muscle tissue of large mammals (21) and between 4 and 5 to 1 in human red cells (22). The ethanolamine phosphoglycerides from brain gray matter of all mammalian species we have so far studied (13) have some 18 to 26% of these fatty acids as docosahexaenoic acid (C22:6, $n - 3$). In view of such evidence, it would be necessary to prove that linolenic acid was nonessential in man before adopting a nutrition policy on the assumption that it was not required.

3. *Comparative:* The bulk of the evidence on the requirements for structural lipids is derived from experiments with vegetable oils and laboratory rats. Rat lipids are responsive to changes in dietary fats. In the rat, the chain-elongation and desaturation process seems to be rapid, and large amounts of elongation products are found in the tissue phosphoglycerides. Unfortunately, the conditions which apply to the laboratory rat are not those which pertain to a large mammal.

For example, in Fig. 7 we have plotted the EPG fatty acids from rat and guinea pig liver. Although these animals were fed identical diets, there is a marked contrast in the LCP and SCP. The rat liver EPG has considerably more LCP and less SCP than that of the guinea pig.

Finally, in Table 3 we have compared the EPG isolated from colostrum of the human and the cow (*Bos taurus*). Whereas the EPG from the cow has a smaller polyenoic content than the human, it has more linolenic, more eicosapentaenoic, and more docosapentaenoic acids, but very little docosahexaenoic acid. The linolenic, eicosapentaenoic, and docosapentaenoic acids have three carbon-saturated end chains ($n - 3$) and are precursors of docosahexaenoic acid. In other words, despite the presence of significant amounts of precursors, the end product of the $n - 3$ chain elongation and desaturation process is not reached in the cow.

This failure to complete the chain-elongation and desaturation process, despite an obvious availability of parent acids and intermediates, is a common feature of a number of large mammalian species (13), and it must be concluded that the conditions which operate in small and large mammals in this respect are not identical. The fact that significant chain elongation and desaturation does occur in the cow and related species is evidenced by the presence of the enzyme systems and the presence of its products in species living entirely on a plant food structure. In other words, the availability of a precursor does not guarantee complete chain elongation and desaturation, for example, linolenic acid to docosahexaenoic acid. The significance of

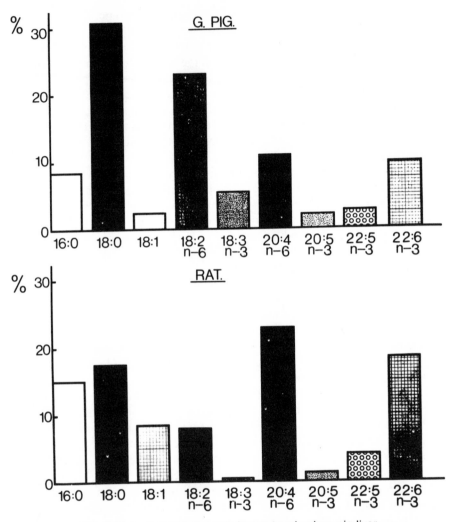

FIG. 7. Plots of EPG fatty acids from rat and guinea pig liver.

this acid is that although other tissues may accumulate n - 3 precursors, the brain of large mammals has been found specifically to contain docosahexaenoic acid.

From this evidence we would conclude that a broad spectrum of lipid nutrients as opposed to the single fatty acid recommended in the "codex" may be desirable for optimum performance as opposed to minimal requirements.

TABLE 3. *Human versus cow colostrum EPG and TG*

Fatty acid	Ethanolamine phosphoglyceride		Triglyceride	
	Human	Cow	Human	Cow
12:0	2.0	1.4	1.5	2.7
14:0	3.0	3.1	5.2	11.0
16:0	12.1	9.4	25.6	29.9
16:1	2.2	0.8	2.7	3.1
18:0	18.0	13.1	8.4	12.9
18:1	19.0	31.1	35.6	29.1
18:2, *n* - 6	10.0	11.0	8.7	2.1
18:3, *n* - 3	1.3	3.9	3.4	1.9
20:2, *n* - 6	0.9	0.2	0.8	0.02
20:3, *n* - 3	2.8	1.3	0.5	0.1
20:4, *n* - 6	12.4	5.9	2.2	0.2
20:5, *n* - 3	0.3	3.8	0.1	0.3
22:4, *n* - 6	2.1	0.1	1.3	0.03
22:5, *n* - 3	3.2	4.9	0.6	0.45
22:6, *n* - 3	7.0	0.6	1.1	0.1

XI. SUMMARY

Thirty-two samples of human milk have been studied and were found to contain polyenoic acid which account for between 9 and 10% of the calories with 2.0% present as the LCP. The ethanolamine phosphoglycerides contained the highest proportions of LCP. The outstanding characteristic of the milk is the wide variation in polyenoate content, both from individual to individual and in the same individual after delivery. We found the colostrum to be particularly rich in arachidonic and docosahexaenoic acids. The milk phosphoglycerides in the first postnatal week lost a substantial proportion of their polyenoic content in a manner which might be expected from a marginal deficiency of essential lipid. Thereafter, the fatty acid pattern recovered but seldom reached the high content of long-chain polyenoic acids found in the colostrum.

Milk triglycerides from Tanzanian mothers contained significantly more linoleic acid than those mothers from London, but such differences are under further investigation.

The sphingomyelin fraction was rich in lignoceric and nervonic acids, but the distribution of the long-chain saturated and monounsaturated acids varied in a manner similar to the polyenoic acids.

It is likely that human milk provides an important source of structural lipid for organogenesis in the neonate, particularly with respect to membrane-rich tissue such as the brain and the vasculature. From what is known from the analytical, comparative, and metabolic studies, a broad spectrum of long-chain acids would contribute qualitatively to the development of the infant.

ACKNOWLEDGMENTS

We are deeply grateful to Professor B. Laurence and the many wives who participated. In particular, we would like to acknowledge the assistance of Mrs. L. Springett and Miss P. Stevens.

This work was supported by a generous grant from the Medical Research Council, no. 971/741/T.

REFERENCES

1. Widdowson, E. M. (1970): *Lancet*, 1:901.
2. Winick, M. (1968): *Paediat. Res.*, 2:352.
3. Widdowson, E. M. (1965): *Lancet*, 2:1099.
4. Tomarelli, R. M., Meyer, B. J., Weaber, J. R., and Bernhar, F. W. (1968): *J. Nutr.*, 95:293.
5. Filer, L. J., Mattson, F. H., and Fomon, S. J. (1969): *J. Nutr.*, 99:293.
6. György, P., and Rose, C. S. (1949): *Ann. N.Y. Acad. Sci.*, 52:231.
7. Hassan, H., Hashim, S. A., Itallie, Van, and Sekrell, W. H. (1966): *Am. J. Clin. Nutr.*, 19:147.
8. Oski, F. A., and Baruess, L. A. (1967): *J. Pediat.*, 70:211.
9. Paoletti, R., and Galli, C. (1972): In: *Lipids, Malnutrition and the Developing Brain*, Ciba Foundation Symposium. Elsevier, Excerpta Medica and North-Holland, Amsterdam.
10. Cravioto, J., Pinero, C., Arroyo, M., and Alcalde, E. (1969): In: *Nutrition in Preschool and School Age* (Swedish Nutr. Fdn. Symp. VII), edited by G. Blix. Almquist & Wiksell, Uppsala, p. 85.
11. Sinclair, A. J., and Crawford, M. A. (1972): *J. Neurochem.*, 19:1753.
12. Sinclair, A. J., and Crawford, M. A. (1972): *Brit. J. Nutr.*, 29:127.
13. Crawford, M. A., and Sinclair, A. J. (1972): In: *Lipids, Malnutrition and the Developing Brain*, Elsevier, Excerpta Medica and North-Holland, Amsterdam.
14. Mohrhauer, H., and Holman, R. T. (1963): *J. Lipid Res.*, 4:151.
15. Fomon, S. J. (1967): *Infant Nutrition*. Saunders, Philadelphia.
16. Insull, W. V., and Ahrens, E. H. (1959): *Biochem. J.*, 72:27.
17. Morrison, W. R. (1968): *Lipids*, 3:101.
18. Morrison, W. R., and Smith, L. M. (1967): *Lipids*, 2:178.
19. Osborn, G. R. (1968): In: *Le Role de la Paroi Arterielle dans l'Atherogenese*. Editions du Centre National de la Recherche Scientifique, Paris, p. 97.
20. Brenner, R. R., and Jose, P. (1965): *J. Nutr.*, 85:196.
21. Crawford, M. A., Gale, M. M., and Woodford, M. H. (1969): *Biochem. J.*, 115:25.
22. Dodge, J. T., and Phillips, G. B. (1967): *J. Lipid Res.*, 8:667.

Dietary Lipids and Postnatal Development
Raven Press, New York © 1973

Absorption of Dietary Lipids and Postnatal Development

Shlomo Eisenberg

Department of Medicine B, Hadassah University Hospital, and The Hebrew-University Hadassah Medical School, Jerusalem, Israel

I. INTRODUCTION

Absorption of dietary lipids through the gastrointestinal tract represents a process of transport of water-insoluble molecules from one water phase, the intestinal lumen, to another water phase, the lymph and plasma. During this process, esterified lipids are hydrolyzed to simple amphipathic molecules, solubilized in luminal water, transferred into intestinal absorptive cells, and secreted to intestinal lymph in the form of lipoproteins. Since many aspects of lipid absorption have been reviewed in the last decade (1–5), I will describe here only briefly the processes occurring until lipids reach the intestinal cells. The recent progress of the understanding of intestinal lipoprotein formation, intracellular transport, and passage to lymph channels as related to postnatal development will be emphasized.

Lipids absorbed from the gastrointestinal tract are derived mainly from diet and some from bile (Table 1). These two sources of lipids are indistinguishable and are metabolically related. Triglycerides constitute the main dietary lipid, about 70 to 100 g/day in adults on regular diet. Phospholipids and cholesterol originate from both diet and bile, and are absorbed during absorptive and postabsorptive periods. Two other lipoid families—con-

TABLE 1. *Lipids absorbed through the gastrointestinal tract*

Lipid	Quantity g/day	Source
Triglycerides	50–150	Diet
Phospholipids	15–30	Diet, bile
Cholesterol	2–3	Diet, bile
Bile salts	10–15	Bile
Vitamins	Trace	Diet

57

jugated bile salts and lipid-soluble vitamins – are derived from bile and diet, respectively. The complex process which results in the absorption of these compounds is best viewed as a series of events occurring at various compartments of the gastrointestinal tract, starting in the stomach and emerging within the intestinal lymphatics.

II. PANCREATIC JUICE

The presence of enzymes secreted by the exocrine portion of the pancreas is essential for normal fat absorption. The sodium bicarbonate content of pancreatic juice also contributes to the basic pH necessary for optimal activity of these enzymes. The more important enzymes are pancreatic lipase (glycerol-ester hydrolase, E.C.3.1.1.3) and pancreatic phospholipase (phosphatidyl acyl hydrolase E.C.3.1.1.4).

Pancreatic lipase. The enzyme catalyzes the hydrolysis of ester bonds at the 1 and 3 position of the glycerol moiety of triglycerides and diglycerides (5–7) (Fig. 1) and is also active against 1-acyl monoglycerides but not against 2-acyl monoglycerides (8, 9). The products of hydrolysis are free fatty acids, 1,2-diglycerides, and 2-monoglycerides. Some glycerol liberated during prolonged incubations is due to isomerization of 2-monoglycerides

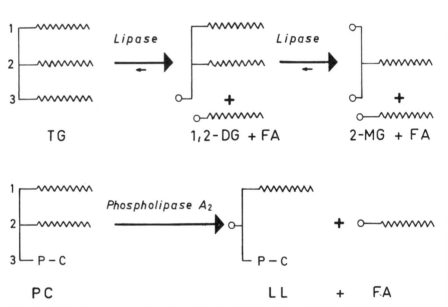

FIG. 1. Schematic presentation of hydrolysis of triglyceride and phosphatidyl choline by pancreatic lipase and phospholipase A_2.

to 1-monoglycerides or 1,2-diglycerides to 1,3-diglycerides and hydrolysis of the ester bonds at the 1 or 3 position of the glycerol molecule (7). Maximal activity of purified enzyme, or fresh pancreatic juice, is at pH of 8.0 to 9.0 and in the presence of 0.1 M NaCl (6, 7). The enzyme has optimal activity against emulsions of triglycerides and only negligible activity toward water-soluble esters (10, 11). It was shown to act at lipid:water interfaces, and the initial velocity of the reaction is proportional to the surface area of the substrate rather than to its concentration (12). A rapid increase of the initial velocity of the reaction has been observed when triacetin, tripropionin, or tributyrin form small aggregates of about 15 molecules each (13). When the effect of fatty acid chain length on the initial velocity of triglyceride hydrolysis by lipase was investigated, it was found that the enzyme is more active against short-chain fatty acids (C_4 to C_8) than against long-chain fatty acids (6). The enzyme also hydrolyzes the ester bond at the 1 position of phosphatidyl choline or phosphatidyl ethanolamine, forming free fatty acid and 2-acyl lysophosphatides (14, 15).

Phospholipase A. The enzyme is specific for the fatty acid at the 2 position of lecithin and phosphatidyl ethanolamine, and is therefore a phospholipase A_2 (Fig. 1) (16, 17). The products of hydrolysis are free fatty acids and 1-acyl lysophosphatides (17). Pancreatic phospholipase is secreted as a proenzyme and is activated *in vitro* by trypsin (18).

Following attack by pancreatic enzymes, nonpolar triglycerides and weakly amphipathic lecithin molecules are converted to strongly amphipathic molecules such as free fatty acids, monoglycerides, and lysophosphatide compounds. In the absence of pancreatic enzymes, fat absorption is largely inhibited, and steatorrhea occurs.

III. CONTRIBUTION OF BILE AND MICELLE FORMATION

Coarse emulsion of fats is produced in the stomach. Fats are liberated from proteins and form droplets of phospholipids and triglycerides following the churning action of the normal gastric motility. However, perhaps the most important function of the stomach in fat absorption is the regulation of delivery of small amounts of the fat emulsion into the duodenum by the pylorus. The presence of fat in the intestine seems to reduce the emptying rate of the stomach and to prevent delivery of an overwhelming quantity of fat into the small intestine (1). Once emulsified fat reaches the duodenum, it is mixed with bile and pancreatic juice already present in the intestinal lumen. Food triglyceride is hydrolized to 2-monoglycerides and free fatty acids which have a limited but important solubility in intestinal luminal water. This solubility is increased 100-fold by the formation of mixed conjugated bile salts–monoglycerides–fatty acids micelles (Fig. 2) (19). The structure and composition of these micelles has been investigated extensively by Hoffman and Borgstrom and has been reviewed previously (20–22). Bile salts, above their critical micellar concentration, form bilayer structures – micelles – with their hydrophilic end toward the water phase and their hydrophobic (and therefore lipophilic) portion in the center of the micelle (21). The shape of the micelles depends on the concentration of the bile salts, and may change from spheres to rods to cylinders and lamellae (21). The intestinal micelles are probably spheres, of an average diameter of 16 to 20 Å (23). The most important property of micellar solution is the ability to solubilize increasing amounts of monoglycerides, free fatty acids, and nonpolar lipid molecules (21). Thus, intestinal lipids are divided into two phases: an oil phase and an aqueous–micellar phase (Fig. 2). Very small quantities of lipids are also soluble in the intestinal water. The lipid composition of the two phases was determined by Hoffman and Borgstrom (19). This study confirmed the expected composition of the phases: the oil phase contained mainly tri- and diglycerides, with some free fatty acids and monoglycerides, whereas the micellar phase contained bile salts, free fatty acids, and monoglycerides, with only traces of tri- and diglycerides (19). The formation of mixed micelles increases the surface area of the lipids by about 100-fold, and greatly enhances their rate of absorption by intestinal absorptive cells. However, bile salts exert several additional effects on fat absorption: (a) In the presence of bile acids, there is a shift

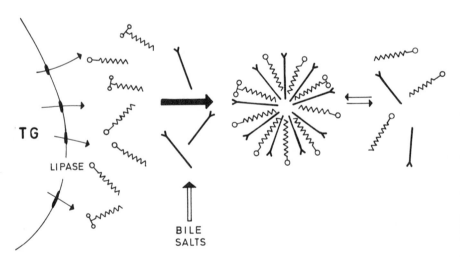

FIG. 2. Schematic presentation of the formation and structure of mixed bile salts–monoglycerides–fatty acids micelles.

of the pH optimum of pancreatic lipase toward 6.0 to 6.5, close to that of the intestine luminal content (24). (b) The high solubility of free fatty acids in micelles shifts the reaction of pancreatic lipolysis from reesterification to hydrolysis (24), as does the continuous passage of lipolytic products into the cells (21). (c) High concentrations of bile salts displace pancreatic lipase from lipid droplets (12), and thus facilitate further hydrolysis of triglycerides by this enzyme. Nevertheless, it should be pointed out that, in the absence of bile, efficient fat absorption can take place, although at a somewhat slower rate (25).

During the normal process of fat absorption and following the interaction of dietary lipids with pancreatic enzymes and bile salts, the intestinal chyme is composed of micelles and small lipid droplets. The surface area of the lipids has increased by several hundred-fold, and they are largely composed of simple amphipathic molecules. They are now ready to be taken up by the intestinal absorptive cells. Most of the lipids are absorbed in the jejunum (26), and most of the bile salts are absorbed in the ileum, reach the liver via the portal circulation, and are resecreted into the bile (27).

IV. ABSORPTION INTO INTESTINAL CELLS

Studies with doubly labeled monoglycerides (28, 29) and lysolecithin (30) have shown that these compounds are absorbed from the intestinal lumen as such. To what extent lecithin, tri-, and diglycerides are absorbed intact, and whether their absorption is of any significance, is unknown. The mechanism of absorption of cholesterol, as free cholesterol, may be different (31). In the rat, more cholesterol is absorbed from the intestine when triolein is added to the diet. When rats were fed mixtures of cholesterol and triolein, the amount of cholesterol absorbed increased with the increase in the amount of cholesterol fed. However, at all concentrations, an almost constant fraction of the cholesterol—0.4—was absorbed, irrespective of the dose fed, with a maximum rate of absorption of about 3 to 4 μmoles/hr (32).

The mode of transfer of lipids across the plasma membrane of the microvilli of absorptive cells was in debate until recently (1–4). Palay and Karlin (33), using electron microscopy, studied morphological changes of intestinal absorptive cells during fat absorption. They described droplets of 500 to 1000 Å in the lumen, between microvilli, and within the cells. Intracellular droplets were enclosed by thin membranes. They concluded that pinocytosis may play an important role in fat absorption. However, it soon became apparent that pinocytosis plays a very minor role in fat absorption, if it functions at all, and that the bulk of the luminal lipids enter the cell by passive diffusion. Strauss and his colleagues (34–37) achieved separation between uptake and esterification of lipids by incubating everted hamster intestinal sacs with a micellar solution of bile salts, monoglycerides, and labeled fatty acids at 0°C for 20 min. Under these conditions, a substantial entry of labeled fatty acids into cells was demonstrated by electron microscopy radio-autography. Ninety-five percent of the label, however, was in the form of free fatty acids, and no lipid droplets were found. When the intestinal sacs were washed and reincubated in a medium without lipids at 37°C, lipid droplets were formed rapidly, and the label was then mainly in esterified lipids. This study demonstrated that the main route of entrance of lipids into the cells is by passive diffusion of molecules, or small micelles, and that their reesterification occurs only later, as predicted by Hoffman and Borgstrom (38) and by Johnston and Borgstrom (39). Using different approaches, Cardell and his co-workers (40) came to the same conclusions for fat absorption in rats as had Rubin and his associates earlier in the human (2, 41).

V. FORMATION OF LIPOPROTEINS

Lipids absorbed into intestinal mucosal cells are released to lymph in the form of lipoprotein particles. The only exception is short-chain fatty acids (C_{12} or less), which are transported via the portal venous blood (42). Chylomicrons and very low density lipoprotein (VLDL), two lipoprotein families synthesized in intestinal cells, are involved in the transport of triglycerides (43). High-density lipoprotein is also synthesized by the intestine, at least in the rat (44), but its significance in fat absorption is as yet unknown.

Very low density lipoprotein and chylomicrons were recently isolated from human intestine, and their size and lipid composition was found to be similar to that of the plasma and lymph lipoproteins (Fig. 3, Table 2). Very low density lipoprotein is composed of 85 to 90% lipids and 5 to 15% protein; up to 99% of chylomicron mass is lipid. Triglycerides constitute 60 to 70% of VLDL lipids and 85 to 95% of chylomicron lipids. Phospholipids and cholesterol are the other two major lipids of either VLDL or chylomicrons (43). The protein moiety of VLDL has been recently shown to be composed of several different apolipoproteins. One apolipoprotein constitutes about 40% of the total VLDL proteins, is identical to the apoprotein of plasma low-density lipoprotein, and is designated apoLDL (45, 46). Another group of small molecular weight proteins [molecular weight 10,000 or less (47)] are collectively designated apoC (48) and constitute about 50% of apoVLDL (45–47). VLDL may also contain small amounts of the major apoproteins of plasma high-density lipoprotein (49).

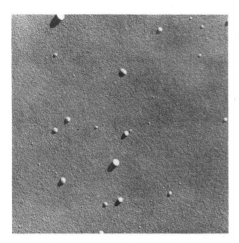 

FIG. 3. Shadow-casted lipoprotein particles extracted from human jejunal biopsies. Left. Particles of size similar to plasma very low density lipoprotein. Right. Particle of size similar to chylomicrons. ×30,375. [From G. N. Tytgat, C. E. Rubin, and D. R. Saunders, (1971): *J. Clin. Invest.,* 50:2065. Reproduced by permission of the authors and the *Journal of Clinical Investigation.*]

The protein moiety of chylomicrons has not been studied in detail so far. However, immunologically it contains the protein designated apoLDL, as well as several other proteins (43, 50). This apolipoprotein is synthesized *in situ* by intestinal absorptive cells (51, 52). The obligatory role of the apoLDL moiety of chylomicrons and VLDL for the formation of these lipoproteins is dramatically demonstrated in the inherited disease abetalipoproteinemia. In this disease, when apoLDL is not synthesized and is not detected in plasma, neither VLDL nor chylomicrons are formed, and

TABLE 2. *Composition of lipoproteins, percent of weight*

Constituent	Chylomicrons	VLDL
PROTEIN	2	10
apoLDL	20*	40
apoC	70*	50
apoA	10*	+
LIPID	98	90
Triglyceride	90	65
Cholesterol	5	15
Phospholipid	5	20

* Ref. (78).

lipids are not absorbed from the gastrointestinal tract (53). Defective absorption of dietary lipids is also found following administration of inhibitors of protein synthesis (ethionine, cycloheximide, puromycin, etc.) to experimental animals (54–57). Thus, although a minor component of chylomicrons and VLDL, normal synthesis of apolipoproteins is a prerequisite for normal fat absorption.

The intracellular site of lipoprotein formation is probably the endoplasmic reticulum, a site where active triglyceride, phospholipid, cholesterol, and protein synthesis occurs. Most of the lipoprotein triglyceride fatty acids originate from absorbed lipids. Two pathways of triglyceride formation are present in the intestine: acylation of monoglycerides and the glycerophosphate–phosphatidic acid–triglyceride pathway. The activity of enzymes involved with these pathways has been demonstrated many times in homogenates of small intestine, and includes acyl-CoA ligase. glycero-kinase, acyl-CoA L-glycerol-3-phosphate acyl transferase, phosphatidate phosphohydrolase, and acyl-CoA; monoglyceride and acyl-CoA diglyceride acyl transferase. Intestinal cells also contain a potent monoglyceride hydrolase activity (1, 4). Under normal conditions, most lymph triglycerides are derived from esterification of 2-monoglycerides and thiol-esters of fatty acids (58). In contrast, a large proportion of lymph lecithin is derived from newly synthesized molecules (59), and only about 40% from acylation of lysolecithin (30). The cholesterol moiety of intestinal lipoproteins is also derived from two sources: cholesterol absorbed from the lumen and cholesterol synthesized *in situ* (60, 61). The contribution of the latter source of cholesterol may be as much as one-third of the total lymph cholesterol (60–63). Of the protein moiety of lipoproteins, the apoLDL protein, at least, is synthesized *in situ* (51).

Chylomicrons are formed only following a fatty meal; VLDL is formed during both absorptive and fasting periods (63), and can be identified in the intestinal cells during the two periods (64). The major source of VLDL lipids during postabsorptive periods is bile lipids; bile diversion or cholestyramine administration greatly diminishes its production (62) and abundance (25) in mucosal cells.

Whether intestinal lipoprotein production may be regulated at a subcellular level is not known, nor is the metabolic relationship of various lipoprotein constituents and the mechanism of packaging lipid and protein into one particle. Also, the relationships between intestinal VLDL and chylomicrons are not clear. Tytgat et al. regard both particles as one family of lipoproteins and hypothesize that the VLDL particles, present in the fasting state, may become chylomicrons during active fat absorption due to addition of triglycerides (64). Ockner et al. argue that intestinal VLDL represents particles which are, at least in part, independent of chylomicron metabolism, and stress its possible role in cholesterol transport (62, 63).

VI. INTRACELLULAR TRANSPORT AND SECRETION OF LIPOPROTEINS

The intracellular location of lipoproteins in intestinal cells and their possible routes of transport have been studied mainly using the electron microscope (2, 64) and electron microscopy radio-autography (65–68). The latter studies, performed in experimental animals, demonstrated that labeled precursors were incorporated into lipid droplets in the endoplasmic reticulum as early as 1 to 2 min after the beginning of incubation. Radio-autographic reaction increased over lipid droplets in the Golgi cisternae during the next 2 to 20 min of incubation, and appeared above intercellular lipid droplets 5 min after introduction of labeled fatty acids into the lumen (65–67). Concentration of label above the Golgi apparatus following pulse-chase labeling of intestinal lipoproteins with ^{3}H-oleic acid was also reported by Dremer (68). Radio-autographic reaction was always more abundant in the apical part of the cell than in its basal part. These studies suggest that lipoproteins, formed in the endoplasmic reticulum, are then transported to the Golgi cisternae, which are in direct communication with the smooth endoplasmic reticulum and are secreted from the Golgi cisternae to intercellular spaces. These lipoproteins were recently isolated from the Golgi apparatus of rat intestine (69).

The morphological features of fat absorption in humans were studied by Rubin and associates during absorptive and postabsorptive phases (2, 64). Lipoprotein particles were identified within absorptive cells, at intercellular spaces, lamina propria, and lymph lacteals. Within cells, single and rows of lipoprotein particles were visualized within profiles of smooth endoplasmic reticulum, especially in the apical part of the cells. Occasionally, lipoprotein particles were identified at areas connecting rough and smooth endoplasmic reticulum structures. Lipoproteins were abundant within Golgi cisternae (Fig. 4), and were found there even when they were not identified elsewhere. Lipoprotein particles were also found within vesicles and vacuoles along the lateral and basal borders of the cells, which were occasionally connected with smooth endoplasmic reticulum formations. Some of these vacuoles, containing lipoprotein particles, appeared to be connected with extracellular spaces (Fig. 4). Many particles were present in intercellular spaces, mainly between the lower portions of adjacent cells and between the basal mem-

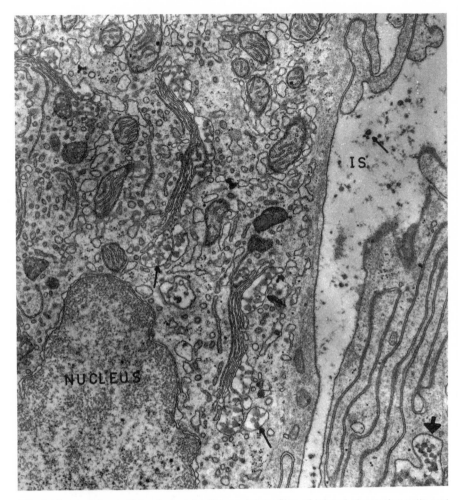

FIG. 4. Lipoprotein particles (arrows) within golgi profiles of jejunal absorptive cells and intercellular spaces (IS). At right lower corner a structural configuration suggesting exit of lipoprotein particles is identified by reverse micropinocytosis (show large arrow). ×19,800. (Unpublished micrograph, from data of G. N. Tytgat and C. E. Rubin, reproduced by courtesy of the authors.)

brane and basal lamina. Lipoprotein particles were found in clusters around gaps in the basal lamina, presumably in transport from intercellular spaces to the lamina propria (Fig. 5). On morphological grounds, the particles could reach the lacteal lumen either through gaps between endothelial cells, or through endothelial cells, or through both.

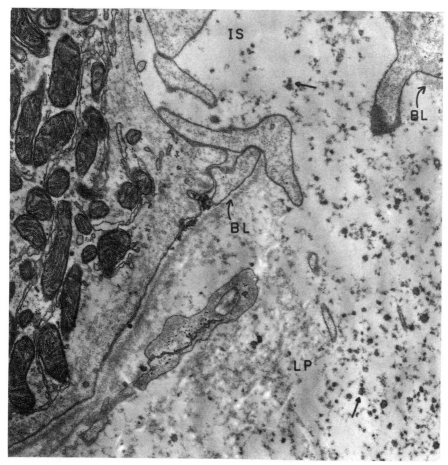

FIG. 5. Lipoprotein particles (arrows) in intercellular space (IS), a gap in the basal lamina (BL), and lamina propria (LP) extracellular space. (Unpublished micrograph from data of G. N. Tytgat and C. E. Rubin, reproduced by courtesy of the authors.)

These findings were interpreted as suggestive of the following sequence: Lipoproteins, formed in the endoplasmic reticulum, are transferred to Golgi cisternae, and leave the cells through the lateral and basal cell membranes by reverse pinocytosis. They then enter the intercellular spaces and reach the lamina propria by passing through gaps in the basal lamina of the absorptive cells. They then begin their relatively long extracellular journey to the lacteal lumen. The same sequence seems to take place during both absorptive and postabsorptive periods.

VII. ABSORPTION OF DIETARY LIPIDS AND POSTNATAL DEVELOPMENT

Relatively little is known about the physiology of fat absorption during neonatal and postnatal periods. The overall process of absorption of dietary lipids and the formation of chylomicrons takes place probably from birth. Hyperchylomicronemia with eruptive xanthomas may be seen in infants with type I hyperlipoproteinemia (70). In this disorder the enzyme system lipoprotein lipase is absent, and chylomicrons are not cleared from circulation.

Bile salts are synthesized by the liver *in utero,* and are found in the gall bladder and intestinal lumen (71, 72). A study on the metabolism of cholic and taurocholic acids in dogs established normal routes of metabolism of these bile acids in the dog fetus 1 week before delivery (73, 74). Moreover, the study also demonstrated normal functioning gall bladder and enterohepatic circulation, the only difference from adult dogs being a hypothesized shunting of some bile acids from portal to systemic circulation.

The development and function of the exocrine pancreas of neonatal rats were reported recently (75). Lipase activity was detected from birth, as were chymotrypsinogen, trypsinogen, and amylase activities. During the first week, only small changes in pancreatic weight and DNA and RNA content were observed (Fig. 6). Between days 6 and 27, pancreatic weight and DNA and RNA content increased rapidly. Lipase activity decreased immediately after birth and increased from day 6, in parallel to pancreatic weight (Fig. 6). Fat-rich diet caused a further increase of pancreatic lipase activity.

Thus, from birth, conditions exist in the intestinal lumen to form simple amphipathic lipids and lipid micelles, similar to those described in adults.

Cellular events of fat absorption in neonates also seem to be similar to those of adults. In a study reported by Rubin in 1966 (2), the morphology of fat absorption in newborn puppies fed colostrum and corn oil was investigated by electron microscopy. The details of fat absorption in these newborn puppies were essentially similar to those reported by Rubin as occurring in the adult, and chylomicron formation seemed to be undisturbed. This observation is of great interest, since it has been reported repeatedly (76) and it was also shown by Rubin that protein absorption in newborns may be quite different from that of the adult, and that micropinocytosis may play an important role in the absorption of immunoglobulins and other proteins.

Normal fat absorption in infants, however, is of extreme importance for normal postnatal development and cannot be overemphasized (77).

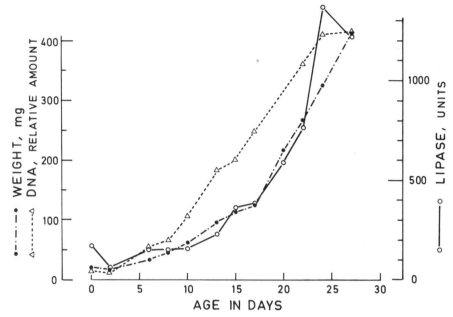

FIG. 6. The changing pattern of pancreatic weight, DNA content, and lipase activity in neonatal rats. [Drawn following data from J. T. Snook (1971): *Am. J. Physiol.*, 221:1388.]

As in the adult, fat malabsorption during the postnatal period may result from defects in any of the events necessary for absorption of dietary lipids described previously (Table 3). These events include absence of bile, malfunction of the pancreas, small bowel diseases, failure to produce VLDL and chylomicrons (as seen in the rare congenital disorder abetalipoproteinemia (53), and obstruction of intestinal lymph channels.

TABLE 3. *Mechanisms of fat malabsorption*

Condition	Defect
Pancreatic insufficiency	Hydrolysis of TG
Biliary atresia	Micelles formation
Small bowel diseases	Absorption of lipids
Abetalipoproteinemia	Formation of lipoproteins
Lymphagiectasis	Secretion of lipoproteins

The clinical consequences of fat malabsorption during infancy, and their influence on postnatal development, may be disastrous. Symptoms include the dramatic clinical picture of steatorrhea and symptoms specific for each disease state. Characteristically, one observes failure to thrive, caloric and protein malnutrition, and multiple vitamin deficiency. If undiagnosed, or untreated, death may supervene.

REFERENCES

1. Senior, J. R. (1966): *J. Lipid Res.*, 5:495.
2. Rubin, C. E. (1966): *Gastroenterology*, 50:65.
3. Dawson, A. M. (1967): *Brit. Med. J.*, 23:247.
4. Johnston, J. M. (1970): In: *Comprehensive Biochemistry*, 18:1.
5. Desnuelle, P. (1971): *Biochimie*, 53:8.
6. Desnuelle, P. (1961): *Adv. Enzymol.*, 23:129.
7. Entressangles, B., Sari, H., and Desnuelle, P. (1966): *Biochim. Biophys. Acta*, 125:600.
8. Mattson, F. H., Benedict, J. H., Martin, J. B., and Beck, L. W. (1952): *J. Nutr.*, 48:335.
9. Hoffman, A. F., and Borgstrom, B. (1963): *Biochim. Biophys. Acta*, 70:317.
10. Sarda, L., and Desnuelle, P. (1958): *Biochim. Biophys. Acta*, 30:513.
11. Brockeroff, H. (1969): *Arch. Biochim. Biophys.*, 134:366.
12. Benzonana, G., and Desnuelle, P. (1965): *Biochim. Biophys. Acta*, 105:121.
13. Entressangles, B., and Desnuelle, P. (1968): *Biochim. Biophys. Acta*, 159:285.
14. De Haas, G. H., Sarda, L., and Roger, J. (1965): *Biochim. Biophys. Acta*, 105:636.
15. Slotboom, A. J., De Haas, G. H., Bonsen, P. P. M., Burbach-Westerhuis, G. J., and Van Deenen, L. L. M. (1970): *Chem. Phys. Lipids*, 4:15.
16. Magee, W. L., Gallai-Hatchard, J., Sanders, H., and Thompson, R. H. S. (1962): *Biochem. J.*, 83:17.
17. De Haas, G. H., Postema, N. M., Nieuwenhuizen, W., and Van Deenen, L. L. M. (1968): *Biochim. Biophys. Acta*, 159:103.
18. De Haas, G. H., Postema, N. M., Nieuwenhuizen, W., and Van Deenen, L. L. M. (1968): *Biochim. Biophys. Acta*, 159:119.
19. Hoffman, A. F., and Borgstrom, B. (1964): *J. Clin. Invest.*, 43:247.
20. Hoffman, A. F. (1966): *Gastroenterology*, 50:56.
21. Hoffman, A. F., and Small, D. M. (1967): *Ann. Rev. Med.*, 18:333.
22. Hoffman, A. F. (1968): In: *Handbook of Physiology*, edited by C. F. Cole. American Physiological Society, Washington, D.C., 5 (Sec. 6), p. 2507.
23. Borgstrom, B. (1965): *Biochim. Biophys. Acta*, 106:171.
24. Borgstrom, B. (1964): *J. Lipid Res.*, 5:522.
25. Porter, H. P., Saunders, D. R., Tytgat, G. N., Brunser, O., and Rubin, C. E. (1971): *Gastroenterology*, 60:1008.
26. Borgstrom, B., Dahlqvist, A., Lundh, G., and Sjovall, J. (1957): *J. Clin. Invest.*, 36:1521.
27. Wilson, J. D. (1968): *J. Lipid Res.*, 9:207.
28. Reiser, R., and Williams, M. C. (1953): *J. Biol. Chem.*, 202:815.
29. Skipski, V. P., Morehouse, M. G., and Deuel, H. J. (1959): *Arch. Biochem. Biophys.*, 81:93.
30. Scow, R. O., Stein, Y., and Stein, O. (1967): *J. Biol. Chem.*, 242:4919.
31. Treadwell, C. R., and Vahonny, G. V. (1968): In: *Handbook of Physiology*, edited by C. F. Cole. American Physiological Society, Washington, D.C., 3 (Sec. 6), p. 1407.
32. Sylven, C., and Borgstrom, B. (1968): *J. Lipid Res.*, 9:596.
33. Palay, S. L., and Karlin, L. J. (1959): *J. Biophys. Biochim. Cytol.*, 5:363, 373.
34. Strauss, E. W., and Ito, S. (1965): *J. Cell Biol.*, 27:101a.
35. Strauss, E. W. (1966): *J. Lipid Res.*, 7:307.
36. Strauss, E. W. (1968): In: *Handbook of Physiology*, edited by C. F. Cole. American Physiological Society, Washington, D.C., 3(Sec. 6), p. 1377.

37. Strauss, E. W., and Arabian, A. A. (1969): *J. Cell Biol.*, 43:140a.
38. Hoffman, A. F., and Borgstrom, B. (1962): *Federation Proc.*, 21:43.
39. Johnston, J. M., and Borgstrom, B. (1964): *Biochim. Biophys. Acta*, 84:412.
40. Cardell, R. R., Badenhausen, S., and Porter, K. R. (1967): *J. Cell Biol.*, 34:123.
41. Phelps, P. C., Rubin, C. E., and Luft, J. H. (1964): *Gastroenterology*, 46:134.
42. Greenberger, N. J., and Skillman, T. G. (1969): *New Engl. J. Med.*, 280:1045.
43. Levy, R. I., Bilheimer, D. W., and Eisenberg, S. (1971): In: *Plasma Lipoproteins*, edited by R. M. S. Smellie. Academic Press, New York.
44. Windmueller, H. G., and Spaeth, A. E. (1972): *J. Lipid Res.*, 13:92.
45. Shore, B., and Shore, V. (1969): *Biochemistry*, 8:4510.
46. Brown, W. V., Levy, R. I., and Fredrickson, D. S. (1969): *J. Biol. Chem.*, 244:5687.
47. Brown, W. V., Levy, R. I., and Fredrickson, D. S. (1970): *Biochim. Biophys. Acta*, 200: 573.
48. Kostner, G., and Alaupovic, P. (1971): *FEBS Letters*, 15:320.
49. Pearlstein, E., Eggena, P., and Aladjem, F. (1971): *Immunochemistry*, 8:865.
50. Rodbell, M., and Fredrickson, D. S. (1959): *J. Biol. Chem.*, 234:562.
51. Windmueller, H. G., and Levy, R. I. (1968): *J. Biol. Chem.*, 243:4878.
52. Kessler, J. I., Stein, J., Dannacker, D., and Narcessian, P. (1970): *J. Biol. Chem.*, 245: 5281.
53. Fredrickson, D. S., Gotto, A. M., and Levy, R. I. (1972): In: *Metabolic Basis of Inherited Disease*, edited by J. B. Stanbury, J. B. Wyngaarten, and D. S. Fredrickson. McGraw-Hill, New York.
54. Sabesin, S. M., and Isselbacher, K. J. (1965): *Science*, 147:1149.
55. Hyams, D. E., Sabesin, S. M., Greenberger, N. J., and Isselbacher, K. J. (1966): *Biochim. Biophys. Acta*, 125:166.
56. Redgrave, T. G. (1966): *Proc. Soc. Exptl. Biol. Med.*, 130:776.
57. Kessler, J. I., Mishkin, S., and Stein, J. (1969): *J. Clin. Invest.*, 48:1397.
58. Mattson, F. H., and Volpenhein, R. A. (1964): *J. Biol. Chem.*, 239:2772.
59. Borgstrom, B. (1952): *Acta Physiol. Scand.*, 25:291.
60. Lindsey, C. A., and Wilson, J. D. (1965): *J. Lipid Res.*, 6:173.
61. Wilson, J. D. (1968): *J. Clin. Invest.*, 47:175.
62. Ockner, R. K., Hughes, F. B., and Isselbacher, K. J. (1969): *J. Clin. Invest.*, 48:2079.
63. Ockner, R. K., Hughes, F. B., and Isselbacher, K. J. (1969): *J. Clin. Invest.*, 48:2367.
64. Tytgat, G. N., Rubin, C. E., and Saunders, D. R. (1971): *J. Clin. Invest.*, 50:2065.
65. Jersild, R. A., Jr. (1966): *Am. J. Anat.*, 118:135.
66. Jersild, R. A., Jr. (1966): *J. Cell Biol.*, 31:413.
67. Jersild, R. A., Jr. (1968): *Anat. Rec.*, 160:217.
68. Dremer, G. B. (1968): *J. Ultrastruct. Res.*, 22:312.
69. Mahley, R. W., Bennett, B. D., Morre, J., Grayd, M. E., Thistlethwaite, W., and LeQuire, V. S. (1971): *Lab. Invest.*, 25:435.
70. Fredrickson, D. S., and Levy, R. I. (1972): In: *Metabolic Basis of Inherited Diseases* (3rd ed.), edited by J. B. Stanbury, J. B. Wyngaarden, and D. S. Fredrickson. McGraw-Hill, New York, Chap. 28.
71. Poley, J. R., Dower, J. C., Owen, C. A., and Strickler, G. B. (1964): *J. Lab. Clin. Med.*, 63:838.
72. Bongiovanni, A. M. (1965): *J. Clin. Endocrinol. Metab.*, 25:678.
73. Jackson, B. T., Smallwood, R. A., Piasecki, G. J., Brown, A. S., Rauschecker, H. F., and Lester, R. (1971): *J. Clin. Invest.*, 50:1286.
74. Smallwood, R. A., Lester, R., Piasecki, G. J., Klein, P. D., Greco, R., and Jackson, B. T. (1972): *J. Clin. Invest.*, 51:1388.
75. Snook, J. T. (1971): *Am. J. Physiol.*, 221:1388.
76. Rodewald, R. (1970): *J. Cell Biol.*, 45:635.
77. Anderson, C. M. (1966): *Arch. Dis. Children*, 41:571.
78. Kostner, G., and Holasek, A. (1972): *Biochemistry*, 11:1217.

Dietary Lipids and Postnatal Development
Raven Press, New York © 1973

Some Recent Findings on Pancreatic Lipase and Colipase

P. Desnuelle

Centre de Biochimie et de Biologie Moléculaire, Marseille, France

I. INTRODUCTION

Dietary long-chain triglycerides are not soluble in water because of their low polarity. Consequently, they never generate true aqueous dispersions in which all solute molecules are relatively independent from the others and surrounded by solvent molecules. Instead, they form plurimolecular particles of varying size separated from the water by a hydrophobic interface in an emulsion. It is noteworthy that these glycerides cannot penetrate the brush border membrane of the small intestine unless they are converted into more polar compounds by hydrolysis. This hydrolysis occurs primarily in the duodenum, and it is catalyzed by pancreatic lipase. The resulting compounds are monoglycerides and free fatty acids which are believed to be absorbed by the mucosa in the form of mixed micelles with bile salts. Monoglycerides and free fatty acids serve for the intracellular resynthesis of triglycerides. Hence, partial hydrolysis of triglycerides by lipase is uniquely required for the passage through the brush border membrane.

Pancreatic lipase is probably the best known of all lipases. In the first part of this chapter some characteristic properties of the enzyme will be discussed briefly. The second part will be devoted to the identification, isolation, and some properties of a pancreatic factor, colipase, which prevents lipase inhibition by the bile salt concentrations normally present in the duodenum during lipolysis.

II. ON SOME PROPERTIES OF PANCREATIC LIPASE

Like other enzymes of the exocrine pancreas, lipase is biosynthesized by the rough membranes of the acinar cells, packed in zymogen granules, and carried to the duodenum by pancreatic juice. Its fundamental properties are related to the fact that, being itself freely soluble in water, it must catalyze the hydrolysis of insoluble substrates at a very high rate.

The most satisfactory technique for the purification of pancreatic lipase was described some years ago by Verger et al. (1). The activity of the purified enzyme was shown to increase with the "concentration" of the emulsion of the triglyceride substrate (2). The concentration of an emulsion is defined as the number of emulsified particles with a given size distribution in 1 liter of emulsion. More precisely, it is related to the "interface concentration" which is itself the area of the interface (expressed in square meters per unit volume of emulsion). After this increase, the activity-concentration plot flattens out, indicating the existence of a maximal rate as in the ordinary Michaelis-Menten treatment. However, in the case of lipase, the correct interpretation (2, 3) is that the enzyme is adsorbed by the plurimolecular substrate particles. When the interface between the particles and water is large enough to accommodate all available lipase molecules, the maximal rate of the lipolysis reaction is attained. For a given weight of triglycerides, the interface concentration is larger when the average size of the particles is smaller. As a consequence, the finer the emulsion, the higher the rate for a given enzyme and substrate concentration (3). Moreover, the Lineweaver-Burk plot derived from the activity-concentration dependence is linear. Hence, an apparent K_m can be defined. This K_m should be expressed in the case of lipase as being the interface concentration of the emulsion giving an experimental rate equal to half the maximal value (3).

In addition, it is of considerable interest that pancreatic lipase does not hydrolyze short-chain glycerides such as triacetin and tripropionin, unless the substrate concentration exceeds the saturation of the solution. In other words, lipase does not recognize isolated molecules dissolved in water, but only aggregates (2). A perfect fit exists in this respect between the enzyme properties and its biological function. Another point of interest is that lipase activity in an isotropic (water-clear) system containing water-soluble glycerides is very much increased at higher ionic strength (4). This increase can be correlated with the formation of micelles. Consequently, lipase does not act only on emulsified particles, but also on micelles of a much smaller size. The smallest micelle size consistent with lipase action is not yet known. However, tripropionin micelles containing not more than 13 monomers are known to be readily attacked by lipase (4).

III. THE PANCREATIC COLIPASE

The two hormonal activities, cholecystokinin, inducing bile secretion, and pancreozymin, promoting the outflow of pancreatic enzymes, are now known to be a property of the same molecule. Consequently, bile and pancreatic enzyme flows can be expected to be synchronized to a certain extent. The result is that lipolysis probably occurs in the duodenum in the presence of relatively high concentrations of bile salts and other bile constituents.

The very complex influence exerted by bile salts on lipolysis cannot be fully discussed in this short presentation. However, it is of interest to note that bile salts have been observed for several years to be strongly inhibitory for lipase in concentrations not distinctly higher than these existing in the duodenum (5). Moreover, it was noticed in the course of the purification of lipase that the purified enzyme was more sensitive to bile salts than the crude one. It could, therefore, be assumed that the cruder preparations contained a factor preventing inhibition and that this factor was removed during lipase purification (6, 7).

The factor purified from crude porcine lipase preparations and porcine pancreatic juice (8) was recently identified as a small protein with a molecular weight not exceeding 10,000 and devoid of methionine and tryptophan. This protein was called colipase. It contains five disulfide bridges, which account for the high stability of the molecule toward heat and extreme pH (8). More recently, another technique for the purification of colipase from porcine pancreas has been published (9).

In our experience, the best technique for colipase purification is first to prepare a fully defatted pancreas powder which is subsequently extracted. A final delipidation of the extract is obtained with the aid of a partition between *n*-butanol and an aqueous ammonium sulfate solution. Then the colipase is purified by two chromatographies on QAE and SP Sephadex. The QAE Sephadex separates two peaks with colipase activity. It is not yet known whether the second colipase preexists in porcine pancreas or is artifactually formed from the first during the purification.

Addition of colipase to colipase-free lipase increases the activity of the latter by a factor not exceeding 1.5. As already reported earlier, the real function of colipase is to keep lipase active in the presence of bile salts. For instance, colipase-free lipase is almost completely inhibited by 2 mM taurodesoxycholate. Addition to the system of about 2 moles of colipase for 1 mole of lipase restores normal activity in this system.

The mechanism of colipase function will not be ascertained as long as the origin of the inhibitory effect of bile salts remains unclear. It may be pointed out, however, that lipase inhibition by taurodesoxycholate in a system at pH 6.0 with emulsified tributyrin as substrate is of the competitive type, and that the inhibition constant is 0.15 mM. This constant goes up to 3.2 mM in the presence of 2 moles/mole of colipase. Hog pancreas appears to contain about 25 mg of colipase. The molar ratio of colipase to lipase in porcine pancreatic juice is approximately 2.

The biosyntheses of lipase and colipase by human pancreatic acinar cells are probably independent of one another. A young girl with a complete congenital lipase deficiency was observed in Marseilles to have a normal level of colipase (C. Figarella, personal communication). Conversely, it can be expected that colipase deficiencies are sometimes erroneously considered to be lipase deficiencies.

REFERENCES

1. Verger, R., de Haas, G. H., Sarda, L., and Desnuelle, P. (1969): *Biochim. Biophys. Acta,* 188:272.
2. Sarda, L., and Desnuelle, P. (1958): *Biochim. Biophys. Acta,* 30:513.
3. Benzonana, G., and Desnuelle, P. (1965): *Biochim. Biophys. Acta,* 105:121.
4. Entressangles, B., and Desnuelle, P. (1968): *Biochim. Biophys. Acta,* 159:285.
5. Benzonana, G., and Desnuelle, P. (1968): *Biochim. Biophys. Acta,* 164:47.
6. Baskys, B., Klein, E., and Leur, F. W. (1963): *Arch. Biochem. Biophys.,* 102:201.
7. Morgan, R. G. H., Barrowman, J., and Börgstrom, B. (1969): *Biochim. Biophys. Acta,* 175:65.
8. Maylié, M. F., Charles, M., Gache, C., and Desnuelle, P. (1971): *Biochim. Biophys. Acta,* 229:286.
9. Erlanson, C., and Borgström, B. (1972): *Biochim. Biophys. Acta,* 271:400.

Dietary Lipids and Postnatal Development
Raven Press, New York © 1973

Studies on the Gastric Lipolysis of Milk Lipids in Suckling Rats and in Human Infants

Thomas Olivecrona, Olle Hernell, Torbjörn Egelrud,
Åke Billström*, Herbert Helander*,
Gösta Samuelson**, and Bo Fredrikzon**

*Departments of Physiological Chemistry, *Anatomy, and **Pediatrics, University of Umeå, Umeå, Sweden*

During early postnatal life, the demands on the mechanisms for lipid absorption are high. Not only is food intake high in relation to body weight, but the lipid content of the diet is also high. In the rat, lipid accounts for about 80% of the calories in milk, in the human for about 50%. In contrast to this high demand on the mechanisms for lipid digestion and absorption, the intestinal lipolytic mechanism may have a rather low activity in the newborn (1).

I. INTESTINAL LIPID DIGESTION AND ABSORPTION

Pancreatic juice contains two enzymes that are active against neutral lipids (2, 3). The so-called pancreatic lipase (E.C. 3.1.1.3) has much higher activity against insoluble, emulsified substrates than against soluble substrates (2, 4). Its pH optimum is about 9. It is inhibited by bile salts in the concentrations normally found in the small intestine. However, a small protein also present in pancreatic juice, the so-called colipase, restores the activity in the presence of bile salts but with a pH optimum of about 6 (5, 6). The main products of the action of this lipase are monoglycerides and free fatty acids. Together with the bile salts, these products form mixed micelles which carry the lipids to the intestinal mucosa for absorption (7). Another lipase in the pancreatic juice, sometimes called the pancreatic carboxyl esterase (E.C. 3.1.1.1), is more active against micellar or soluble substrates than against insoluble, emulsified substrates (2, 8). In contrast to the classical lipase, it is strongly stimulated by bile salts (8).

The few studies that have been performed on the lipases in human infants or in experimental animals have usually not differentiated these two enzymes. Recent studies in newborn human infants (9) showed that the total lipase activity in intestinal contents is sometimes quite low, and that the bile salt concentration is almost always low (0.4 to 1.5 mmole/liter as compared to 2.5 to 10 mmole/liter in the adult). Bradshaw and Rutter (10) found that the total lipase activity in the pancreas of the rat, as measured in the presence of bile salt and at pH 8.5, was already high at birth. However, during the suckling period, most of the lipase activity was accounted for by the bile salt-stimulated lipase, the esterase, whereas the activity of the classical lipase was quite low. The actual hydrolytic activity against triglyceride in the intestine at pH 6 to 7 may be rather low during the suckling period. We have therefore explored the possibility that lipolytic systems other than those dependent on the pancreatic enzymes are important in the newborn.

II. GASTRIC LIPOLYSIS

We first considered the possibility that lipid digestion might occur in the stomach as well as in the small intestine. A lipase with an acid pH optimum has been demonstrated in gastric contents of several species (for review, see 11). In a suckling rat the milk is clotted upon entering the stomach and forms a semisolid coagel (Fig. 1). Such rats always have the stomach almost full of clotted milk. In experiments with 10-day-old suckling rats, we observed that more than half of administered labeled lipids were still in the stomach 1 hr after they had been given perorally. Thus, the milk lipids are exposed to the action of lipases in the stomach for a considerable time before they are passed on to the small intestine.

To see if milk lipids are hydrolyzed to a significant degree in the stomach, we compared the lipid composition of rat milk and of gastric contents of 10-day-old suckling rats (12). The lipid concentration in the gastric contents was higher than in the milk, presumably because of absorption of water (Table 1). The main lipid in milk was triglyceride. In contrast, almost half of the glycerides in the gastric contents were partial glycerides, mainly diglycerides. Free fatty acids were also present in high concentration. There was slightly less free fatty acid than partial glycerides, as if some of the fatty acids that had been liberated upon hydrolysis of the triglycerides to di- and monoglycerides had also been absorbed.

Milk lipids differ from most other animal triglycerides in having a relatively high proportion of short- or medium-chain fatty acids (13). These fatty acids are predominantly located at the 3 position of the sn-glycerol molecule (14, 15). Thus, most triglyceride molecules will have two long-chain and one medium-chain fatty acid. The fatty acid composition of the

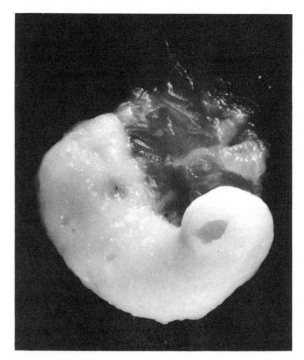

FIG. 1. A stomach from a suckling rat cut open to demonstrate its semisolid content of clotted milk.

TABLE 1. *Lipid composition of rat milk and of gastric contents from 10-day-old rats*

Lipid	Milk	Gastric contents
Triglycerides	59 ± 8.0	127 ± 40
Diglycerides	2.8 ± 0.5	93 ± 18
Monoglycerides	7.4 ± 2.0	16 ± 2
Free fatty acids	0.3 ± 0.1	95 ± 29
Total cholesterol	4.3 ± 0.6	12 ± 2.2
Esterified cholesterol	3.9 ± 0.6	11 ± 2.0
Phospholipids	0.3 ± 0.1	1 ± 0.4

Values are μmole/g wet weight and are mean ± S.D. of 5 to 7 samples.

Each sample of gastric contents represents pooled material from one litter of animals, i.e., about ten rats.

From reference 12 by courtesy of the Williams & Wilkins Company.

triglycerides of the gastric contents was similar to that of the rat milk triglycerides (Table 2). The diglycerides had a lower content of medium-chain fatty acids, and the free fatty acids were mainly of medium-chain length. Thus, the lipolysis in the stomach liberated the medium-chain fatty acid from the 3 position of the sn-triacylglycerol molecule, leaving a diacylglycerol with two long-chain fatty acids. This is more clearly shown when the relative content of medium- and long-chain fatty acids is expressed as the molar ratio of C8 to C12 fatty acids divided by C16 to C18 fatty acids. This ratio was 0.6 in the milk triglycerides, whereas it was only 0.4 in the diglycerides from the gastric contents and as high as 3.4 in the free fatty acids.

TABLE 2. *Fatty acid composition of lipids from rat milk and from gastric contents of 10-day-old rats*

Fatty acid carbon no.	Milk triglycerides	Gastric contents		
		Triglycerides	Diglycerides	Fatty acids
8	3.8 ± 0.8	3.4 ± 1.5	1.6 ± 0.4	8.2 ± 3.7
10	12.0 ± 2.2	14.9 ± 6.7	7.4 ± 0.8	33.4 ± 5.1
12	10.1 ± 1.2	12.7 ± 3.4	9.7 ± 1.1	23.3 ± 2.5
14	10.3 ± 0.7	10.5 ± 1.7	14.5 ± 0.7	6.1 ± 1.0
16	26.6 ± 1.8	22.4 ± 3.7	33.9 ± 0.5	8.7 ± 2.1
18	37.3 ± 3.1	36.1 ± 9.0	32.9 ± 1.9	20.4 ± 6.4

Percentages by weight. Mean ± S.D. of 5 to 7 samples.
Each sample of gastric contents represents pooled material from one litter of animals, i.e., about ten rats.
From reference 12 by courtesy of the Williams & Wilkins Company.

Figure 2 shows the distribution of the fatty acids among the lipid classes. In the milk essentially all the fatty acids were in the triglycerides, whereas in the gastric contents most of the medium-chain fatty acids were present as free fatty acids and the long-chain fatty acids were almost equally distributed between the tri- and diglycerides. This analysis gives only a mean composition of the lipids present in the stomach. The rats were taken directly from suckling, and their stomachs certainly contained some milk that had just been swallowed and had presumably not been hydrolyzed to any significant degree. The lipids passing into the duodenum may well have been considerably more hydrolyzed than the mean composition indicates.

Evidence of intragastric hydrolysis of triglycerides has also been obtained in other species. About one-third of the lipid in the stomach was fatty acid in dogs 4 hr after a fat meal (16), and in humans 10 to 20% of the lipid in

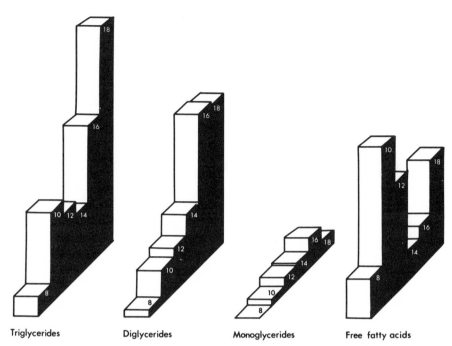

Triglycerides   Diglycerides   Monoglycerides   Free fatty acids

FIG. 2. The distribution of fatty acids between the lipid classes in gastric contents from 10-day-old suckling rats. The molar concentration of each lipid class was multiplied by its molar fatty acid composition. The numbers designate the carbon chain length.

the stomach was present as fatty acid 20 min after a test meal (17). Both these studies concerned adults, and both show that gastric lipolysis occurs not only in the newborn, but at all ages.

In human infants, intestinal content can easily regurgitate into the stomach. Normally it would therefore be difficult to distinguish gastric and intestinal factors in the lipolysis of milk lipids in the infant. However, we have had the opportunity to study three infants with pyloric stenosis. The nature of their disease limits, if not excludes, any interference from regurgitated intestinal content on the gastric lipolysis proper. In all three infants the clinical diagnosis was verified by X-ray examination and was later confirmed at operation. We shall report here the results from only one of the infants, an 11-week-old boy, but the results obtained in the other two infants were entirely consistent.

Gastric content was aspirated by gentle suction through a gastric tube. The infant was then given 30 ml of human milk through a formula bottle. The milk had been heated at 90°C for 5 min, and was at room temperature when given. At intervals, samples of gastric content were taken through the tube and the lipid composition was analyzed as previously described (12). The pH of the samples ranged from 5.6 to 3.1. There was a decrease of the triglycerides and a concomitant increase of partial glycerides in the gastric contents (Fig. 3) with time. As was previously found in suckling rats (12), the partial glycerides were mainly diglycerides. The rate of hydrolysis was about 1 mmole of fatty acid liberated per hour from the milk triglycerides. If all milk triglycerides must be hydrolyzed to monoglycerides and free fatty acids before being absorbed, then the total rate of hydrolysis required would be of the order of 2 to 3 mmoles of fatty acid per hour in

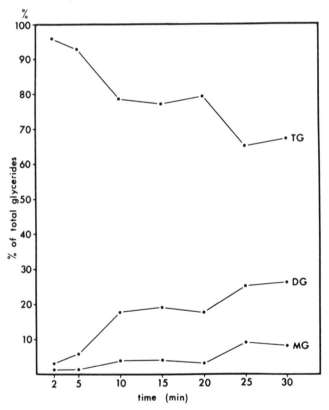

FIG. 3. Composition of lipids in gastric content from a child with pyloric stenosis, at different time intervals after administration of human milk. The lipids were extracted and separated by thin-layer chromatography as previously described (12). Quantification was by determination of glycerol in the lipid fractions (30). TG, DG, and MG are triglycerides, diglycerides, and monoglycerides, respectively.

the human infant. Thus, the gastric lipolysis might account for a sizable proportion of the total hydrolysis. In these experiments, the milk lipase had been inactivated by briefly heating the milk before it was administered. If the milk lipase aids in the gastric lipolysis, as will be discussed below, then the total rate of this process would be further increased.

Thus, the gastric lipolysis may be quantitatively important in the human infant. In addition, it may aid in the dispersion of the lipids for efficient further hydrolysis in the intestine. Native milk fat droplets are not a good substrate for the pancreatic lipases (11). In a model experiment, we incubated rat milk with homogenates of rat tongue (the lipase found in contents of the rat stomach is probably secreted from glands in the back part of the tongue, see below) or of rat pancreas, respectively (Fig. 4). Milk was first

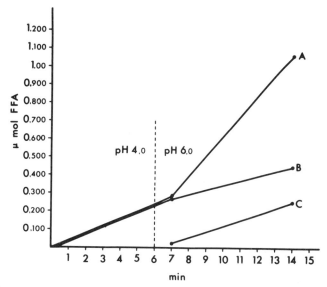

FIG. 4. Effect of preincubation with tongue lipase on hydrolysis of rat milk lipids by pancreatic lipases. Fresh rat milk was incubated in a dilute potassium phosphate buffer also containing bovine serum albumin and 100 mM NaCl. The pH was initially 4, but was adjusted to 6 after 6 min.
Additions were
A: 0 min, tongue homogenate; 7 min, taurocholic acid and pancreatic homogenate.
B: 0 min, tongue homogenate; 7 min, taurocholic acid.
C: 0 min, no addition except buffer to make the same volume as in A and B; 7 min, taurocholic acid and pancreatic homogenate.
The final concentration of taurocholic acid in A, B, and C was 4 mM.
Samples were assayed for fatty acids by the method of Dole (31).
Taurocholic acid made a small contribution to the titration values. The data have been corrected for this contribution.

incubated for 6 min at pH 4, 37°C, with or without tongue homogenate, and pancreatic homogenate and taurocholic acid was then added. The total rate of hydrolysis increased more when the milk had first been exposed to the tongue homogenate. This experiment supports the idea that the gastric lipolysis may change the milk fat droplets so that they are more easily attacked by the intestinal lipolytic mechanisms.

III. THE LIPASES INVOLVED

The results described above demonstrate that lipids are rapidly hydrolyzed in the stomach. Since the gastric acidity itself does not hydrolyze the ester bonds (11), a lipase must be involved. The nature and origin of this enzyme poses some interesting questions.

A. Lipase in Gastric Contents

Several investigators have reported the presence of a lipase in gastric contents (for review, see 11). A lipase has also been extracted from the gastric mucosa (18) and has been demonstrated histochemically in the mucosal cells (19). The most thorough investigation on a lipase from gastric contents was carried out by Cohen, Morgan, and Hofmann (11). They found lipase activity in all samples of gastric juice from 73 subjects. This lipase was stable at pH 2 and had a broad pH optimum between 4 and 7; it was thus both stable and active under conditions prevailing in the stomach during digestion of a meal. This lipase appears to be present also in gastric contents of infants. Figure 5 shows the results from an experiment in which gastric juice from an infant with pyloric stenosis was incubated with an equal volume of human milk. The milk had been briefly heated to inactivate the milk lipase. The pH of the incubation was 4.7. As was observed in the *in vivo* experiments, the milk triglycerides were hydrolyzed, forming mainly diglycerides and free fatty acids.

B. Lingual Lipase and Pregastric Esterase

Although the above data show that there is a potent lipase in gastric contents, they do not tell us from where the enzyme is secreted. Recently, Hamosh and Scow (20) have reported that a lipase is present in the back part of the tongue and in some other pharyngeal tissues in the rat (Table 3). Most of this lipase is in the von Ebner's glands, which are serous glands situated beneath the circumvallate papilla of the tongue, with the ducts emerging into the trough of the papilla (21). This lipase has a pH optimum of about 5 and is stable at pH 3 and 37°C for several hours. It does not seem to

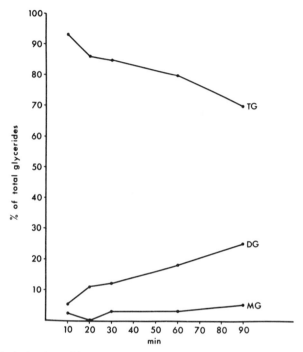

FIG. 5. Hydrolysis *in vitro* of human milk lipids by gastric juice from a child with pyloric stenosis. Equal volumes of human milk and of gastric juice were mixed and incubated at 37°C. The milk had been briefly heated to inactivate the milk lipases. The gastric aspirate from an infant with pyloric stenosis was centrifuged, and the supernatant was used for the incubation. The pH after mixing was 4.7. When milk was incubated under identical conditions but with boiled gastric juice, there was no hydrolysis of the milk lipids. Analysis of the lipids was as in Fig. 3.

TABLE 3. *Properties of lipase from rat tongue*

pH optimum	4.5–5.4
Activity per g of von Ebner's gland[a]	50,000
Inactivation at pH 3 and 37°C	Little
Activators needed	None
Main products formed from triglyceride	Diglyceride and fatty acid

[a]Micromoles fatty acid released from triglyceride per hour and g wet tissue.

Based on the work of Hamosh and Scow (20); with the permission of the authors, except for the information on acid inactivation which is from unpublished work by Dr. Åke Billström.

require any activator and hydrolyzes both long- and short-chain triglyce-rides. Hamosh and Scow demonstrated that the main products formed when this enzyme acts *in vitro* on long-chain triglycerides are diglycerides and free fatty acids. It is most likely, therefore, that this is the enzyme respon-sible for the lipolysis observed in the stomach of the suckling rats.

In calves also, a lipase is secreted from the tissues in the pharyngeal region (22), but the properties of this lipase have not been investigated in detail. When a milk meal was fed perorally to calves, 48% of the butyric acid and 16% of the long-chain fatty acids were released as free fatty acids before the lipid passed into the duodenum (23). In contrast, when the milk was infused directly into the stomach (abomasum) and the lipase-containing tissues in the pharyngeal region were thus circumvented, only 10% of the butyric acid and only 4% of the long-chain fatty acids were released before passage into the duodenum. The implication is that the milk normally becomes mixed with the lipase already in the pharynx before it enters the stomach and is clotted by the rennet. An indication that this enzyme may be particularly important in the early postnatal period is that its activity decreases markedly with age of the calves (22). This is in contrast to the rat, in which the enzyme activity is higher in the tongue of the adult animal than in the infant (20).

In man we do not yet know from where the lipase is secreted.

C. Milk Lipase

Milk from many species contains lipases (24–26). It would indeed be an interesting system if the intestinal mechanisms for lipid digestion were supplemented by an enzyme in the milk. In the rat there is virtually no lipolytic activity in the milk (27). In cow's milk the main enzyme is a lipo-protein lipase. We have recently purified this enzyme (28); its low stability and its substrate specificity makes it unlikely that this enzyme plays a major role in the digestion of the milk lipids.

In human milk, however, there is a lipase which may well play a role in the digestion of the milk lipids (24). Actually, human milk contains two lipases (Table 4). One is a lipoprotein lipase similar to the one in bovine milk. This enzyme has little activity against milk lipids even if a serum lipoprotein activator is added. Furthermore, the pH optimum is above 8, and the enzyme has little activity at pH 7 or below. Thus, this enzyme probably never has any significant activity against the milk lipids *in vivo*. The other enzyme is found almost exclusively in skim milk. It has virtually no activity against the milk lipids in fresh milk, but is activated by bile salts. In the presence of bile salt and at neutral pH, the activity of this enzyme is

TABLE 4. *Lipases in human milk*

	Serum-activated milk lipase	Bile salt-activated milk lipase
Cream	++++	−
Skim milk	+	++++
Addition of serum	Stimulates	Inhibits
Addition of heparin	Stimulates	−
1 M NaCl	Inhibits	Inhibits
Addition of bile salts	Inhibits	Stimulates
Activity per ml milk[a]	<1	37–44
Inactivation at pH 3.5 and 37°C	Rapid	Little
Inactivation by trypsin[b]	−	Little

[a] Micromoles fatty acid released from triglyceride per hour under suitable conditions for the respective enzymes.

[b] In the presence of bile salt. In the absence of bile salt the enzyme activity is rapidly lost upon incubation with trypsin.

Based on unpublished experiments by Dr. Olle Hernell.

enough to hydrolyze almost completely the lipids in the milk in about 2 hr. It is rather stable at pH above 3.5, but has little enzymatic activity below pH 5. This enzyme probably does not act in the stomach, but is stable enough to survive the acid conditions and may act on the milk lipids when they have passed into the duodenum. Actually, in the human infant the milk often passes quite rapidly into the intestine (29).

IV. GENERAL CONCLUSIONS

Available data, although fragmentary, suggest that the normal mechanism of lipid digestion may not be fully developed at birth. In particular, the activity of the classical pancreatic lipase is probably low, as is the concentration of bile acids in intestinal contents. However, in the stomach there is present a potent lipase which hydrolyzes the lipids, mainly to diglycerides and free fatty acids. This hydrolysis has been demonstrated *in vivo* both in rats and in humans. In the rat the lipase is secreted by tissues in the pharyngeal region, and is probably mixed efficiently into the milk during swallowing and before the milk is clotted in the stomach. We do not yet know which tissue secretes the enzyme in the human. The rates of lipolysis observed *in vivo* in human infants were such that they could account for as much as one-third of the total lipolysis. Furthermore, the action of this lipase transforms the milk fat droplets to a substrate that is readily attacked by the pancreatic lipases, whereas native milk fat is not a good substrate for these enzymes. Thus, the gastric lipolysis facilitates the further digestion of the milk lipids.

Human milk contains, in addition to a lipoprotein lipase, a lipase which may hydrolyze the milk lipids under the conditions prevailing in the intestine. This lipase is not active against the milk lipids in milk, but requires bile salts for activation. It is relatively stable at pH as low as 3.5, and may well survive the acid conditions in the stomach, follow the milk lipids into the intestine, and act there. Milk contains enough of this lipase to catalyze an almost complete hydrolysis of its triglycerides in about 2 hr.

ACKNOWLEDGMENT

This work was supported by the Swedish Medical Research Council (B-13X-727 and B-12X-2298) and the Swedish Nutrition Foundation.

REFERENCES

1. Koldovsky, O. (1969): *Development of the Functions of the Small Intestine in Mammals and Man.* S. Karger, Basel.
2. Sarda, L., and Desnuelle, P. (1958): *Biochim. Biophys. Acta,* 30:513.
3. Morgan, R. G. H., Barrowman, J., Filipek-Wender, H., and Borgström, B. (1968): *Biochim. Biophys. Acta,* 167:355.
4. Desnuelle, P. (1971): *Biochemie,* 53:841.
5. Maylié, M. F., Charles, M., Gache, C., and Desnuelle, P. (1971): *Biochim. Biophys. Acta,* 229:286.
6. Borgström, B., and Erlanson, C. (1971): *Biochim. Biophys. Acta,* 242:509.
7. Hofmann, A. F., and Borgström, B. (1962): *Fed. Proc.,* 21:43.
8. Erlanson, C. (*in press*): *Biochim. Biophys. Acta.*
9. Norman, A., Strandvik, B., and Ojamäe, Ö. (1972): *Acta Paediat. Scand.,* 61:571.
10. Bradshaw, W. S., and Rutter, W. J. (1972): *Biochemistry,* 11:1517.
11. Cohen, M. M. B., Morgan, R. G. H., and Hofmann, A. F. (1971): *Gastroenterology,* 60:1.
12. Helander, H. F., and Olivecrona, T. (1970): *Gastroenterology,* 59:22.
13. Scow, R. O., Mendelson, C. L., Zinder, O., Hamosh, M., and Blanchette-Mackie, E. J. (1973): In: *This Volume.*
14. Breckenridge, W. C., and Kuksis, A. (1968): *Lipids,* 3:291.
15. Breckenridge, W. C., and Kuksis, A. (1968): *Lipids,* 4:197.
16. Douglas, G. J., Reinauer, A. J., Brooks, W. C., and Pratt, J. H. (1953): *Gastroenterology,* 23:452.
17. Borgström, B., Dahlqvist, A., Lundh, G., and Sjövall, J. (1957): *J. Clin. Invest.,* 36:1536.
18. Clark, S. B., Brause, B., and Holt, P. R. (1969): *Gastroenterology,* 56:214.
19. Barrowman, J. A., and Darnton, S. J. (1970): *Gastroenterology,* 59:13.
20. Hamosh, M., and Scow, R. O. (*In Press*): *J. Clin. Invest.*
21. Hand, A. R. (1970): *J. Cell Biol.,* 44:340.
22. Ramsey, H. A., Wise, G. H., and Tove, S. B. (1956): *J. Dairy Sci.,* 39:1312.
23. Otterby, D. E., Ramsey, H. A., and Wise, G. H. (1964): *J. Dairy Sci.,* 47:993.
24. Freundenberg, E. (1953): *Jahrbuch für Kinderheilkunde,* Fasc. 54:1.

25. Chandan, R. C., and Shahani, K. M. (1964): *J. Dairy Sci.,* 47:471.
26. McBride, O. W., and Korn, E. D. (1963): *J. Lipid Res.,* 4:17.
27. Hamosh, M., and Scow, R. O. (1971): *Biochim. Biophys. Acta,* 231:283.
28. Egelrud, T., and Olivecrona, T. (1972): *J. Biol. Chem.,* 247:6212.
29. Cavell, B. (1971): *Acta Paediat. Scand.,* 60:370.
30. Carlson, L. A. (1963): *J. Atheroscl. Res.,* 3:334.
31. Dole, V. P., and Meinertz, H. (1960): *J. Biol. Chem.,* 235:2595.

Dietary Lipids and Postnatal Development
Raven Press, New York © 1973

Role of Lipoprotein Lipase in the Delivery of Dietary Fatty Acids to Lactating Mammary Tissue

Robert O. Scow, Carole R. Mendelson, Oren Zinder,
Margit Hamosh, and E. Joan Blanchette-Mackie

*Section on Endocrinology, Laboratory of Nutrition and Endocrinology, National Institute
of Arthritis, Metabolism and Digestive Diseases, Bethesda, Maryland 20014*

I. ORIGIN OF FATTY ACIDS SECRETED IN MILK

Fatty acids in the form of triglyceride account for more than 95% of the lipid secreted in milk (1). The kinds and amounts of fatty acids present, however, vary with species (2) and with diet (3, 4). Some of the differences seen between species are shown in Fig. 1. Long-chain fatty acids, C_{14} to C_{18} in chain length, accounted for 93% of the lipid in human milk (2), 80% in cow's (2) and rat's milk (4), 65% in goat's milk (2), and only 20% in rabbit's milk (5). Medium-chain fatty acids, C_8 to C_{12}, comprised less than 5% of the lipid in human and cow's milk, 20% in goat's and rat's milk, and 76% in rabbit's milk. Short-chain fatty acids, C_4 and C_6, were present in cow's, goat's, and rabbit's milk, accounting for about 10% of the lipid, but not in human or rat's milk.

The effect of diet on the fatty acid composition of milk lipid has been studied in the human by Insull et al. (3). In their study, shown in Fig. 2, the fat content of the diet and the caloric intake were both varied. The diet contained either 40% lard, 40% corn oil, or no fat. The fatty acid composition of the diet is depicted by stippled bars, serum triglyceride by white bars, and milk lipid by black bars. Lard, shown in the top panel, contains proportionally more palmitic (16:0), stearic (18:0), and oleic (18:1) acid, and less linoleic-linolenic (18:2 + 18:3) acid than corn oil.

The effect of dietary fat on the lipid composition of milk lipid is shown in the upper half of Fig. 2. The C_{16} to C_{18} fatty acid composition of the milk was similar to that of the diet when the diet contained either lard or corn oil, and the caloric intake (2800 to 2900 cal/day) was sufficient to maintain a constant body weight (Fig. 2). The C_{12} and C_{14} fatty acid content of the milk

91

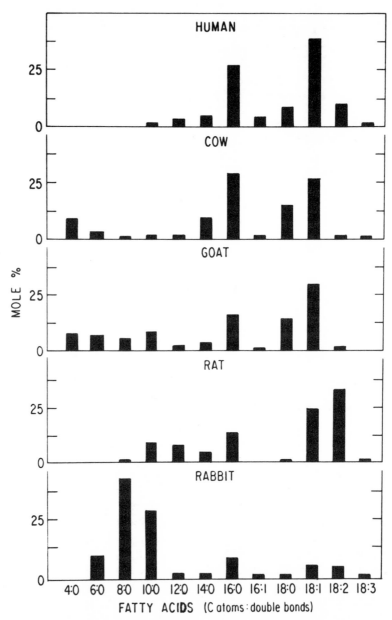

FIG. 1. Fatty acid composition of lipid in milk of several species. Data were taken from Breckenridge and Kuksis (2) for human, cow, and goat; from Beare et al. (4) for rat; and from Carey and Kils (5) for rabbit.

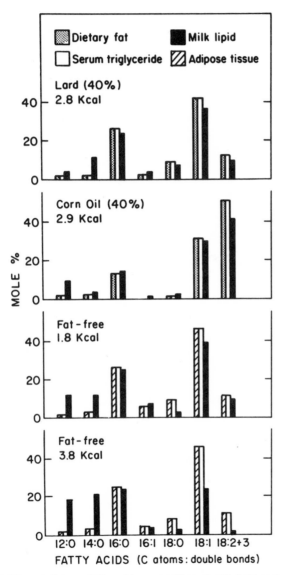

FIG. 2. Effect of different diets on fatty acid composition of milk lipid and serum triglyceride in the human. Data from Insull et al. (3).

lipid, however, was higher than that of the diet and serum triglycerides, suggesting that these fatty acids were synthesized in the mammary gland.

The effect of the fat-free diet on the lipid composition of milk is shown in the lower half of Fig. 2. The fatty acid composition of the adipose tissue is depicted in the figure by striped bars. The composition of the serum triglyceride was similar to that of adipose tissue.

When fat was eliminated from the diet and the caloric intake was reduced to 1800 cal/day, the C_{16} and C_{18} fatty acid composition of milk, except for stearic acid, was similar to that of adipose tissue. The relative amounts of C_{12} and C_{14} fatty acids, however, were increased, and thus accounted for 20% of the fatty acids in the milk.

The fatty acid composition of milk was markedly changed when the caloric intake was increased to 3800 cal/day. The unsaturated C_{16} and C_{18} fatty acids were decreased more than 30%, whereas the C_{12} and C_{14} fatty acids were increased 100%. The latter accounted for 40% of the fatty acids in milk. The C_{10} fatty acid content of the milk was less than 3%. The increased levels of C_{12} and C_{14} fatty acids in the milk and the low levels of these fatty acids in the serum suggest that fatty acid synthesis in mammary tissue is stimulated by a high-caloric, fat-free dietary regimen. These studies also suggest that most of the C_{16} and C_{18} fatty acids secreted in milk are derived from the diet.

Studies in the lactating goat (6, 7) and cow (8, 9) showed that most of the stearic and oleic acid and about two-thirds of the palmitic acid secreted in milk come from plasma triglyceride, and that the rest of the palmitic and most of the shorter-chain fatty acids, C_4 to C_{14}, are synthesized *de novo* in the mammary gland. Recent studies in the lactating rabbit also showed that mammary tissue synthesizes the C_4 to C_{14} fatty acids secreted in milk (5). The C_{18} fatty acid content of milk in rats fed *ad libitum* decreased from 67 to 24 mole-% and the C_{10} to C_{14} fatty acids increased from 10 to 47 mole-% when the fat content of the diet was reduced from 23% to 3% (4).

These findings indicate that dietary lipid is the primary source of fatty acids secreted in milk. When the fat content of the diet is inadequate, fatty acids are synthesized in the mammary gland or mobilized from adipose tissue for milk secretion.

II. UPTAKE OF BLOOD TRIGLYCERIDE AND LIPOPROTEIN LIPASE

Long-chain fatty acids derived from the diet are transported in blood as chylomicron-triglyceride (10, 11). They are taken up and utilized by muscle, liver, adipose tissue, and other organs (12–14). There is now considerable evidence that blood triglyceride is hydrolyzed during uptake to FFA and that this process is catalyzed by lipoprotein lipase (11, 15). The enzyme is quickly released to the blood when heparin is injected, suggesting that the enzyme is present in the capillary wall (15). Lipoprotein lipase is found in most tissues that utilize triglyceride, and the level of activity usually reflects the capacity of the tissue to remove lipid from the blood (11, 16).

III. LIPOPROTEIN LIPASE ACTIVITY IN MAMMARY AND ADIPOSE TISSUE

A. In Pregnant and Lactating Animals

Uptake of blood triglyceride by mammary gland is negligible except during lactation (17). In lactating animals, however, mammary tissue removes up to 50% of the triglyceride that enters the gland (7, 9, 17) and utilizes a major part of the extra lipid ingested (18).

Lipoprotein lipase activity in mammary tissue is also increased during lactation (17, 19, 20). In guinea pigs the activity increases a few days before parturition and the level remains high as long as suckling is continued (19, 20). Although food intake is greatly increased during lactation, there is no appreciable gain in body fat stores in the rat (21, 22) and, accordingly, lipoprotein lipase activity in adipose tissue was found to decrease during the last 3 days of pregnancy and on the second or third day of lactation (23).

We have studied in the rat the effect of pregnancy and lactation on lipoprotein lipase activity in both adipose and mammary tissue and, also, on the triglyceride concentration in plasma (Fig. 3). Lipoprotein lipase activity in adipose tissue increased during the second week of pregnancy, fell to very low levels several days before parturition, and remained low throughout the rest of pregnancy and lactation. Lipoprotein lipase activity in mammary tissue, in contrast, was low during the first 20 days of pregnancy, increased several days before parturition, decreased sharply during parturition, and then immediately increased and remained high throughout lactation. Plasma triglyceride concentration increased between the 12th and the 20th days of pregnancy to 3.3 mM, and then decreased 50% during the last 2 days. The concentration increased again during parturition, and then decreased and remained low, less than 1.0 mM, throughout lactation. The development of hyperlipemia during pregnancy while adipose tissue lipoprotein activity was high (Fig. 3) suggests that plasma triglyceride concentration is independent of the level of lipoprotein lipase activity in adipose tissue.

Nonsuckling causes a marked fall in lipoprotein lipase activity in mammary tissue of lactating guinea pigs (19). Therefore, we studied its effects

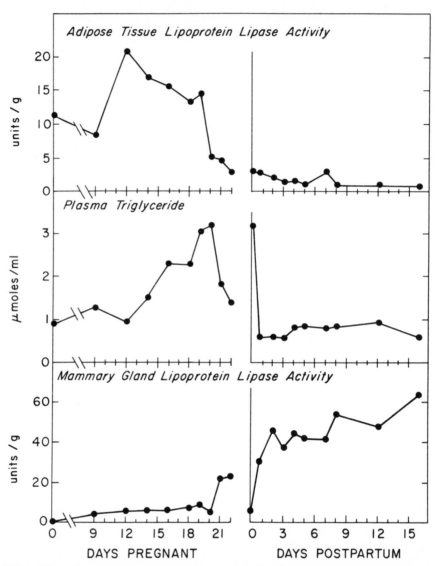

FIG. 3. Effect of pregnancy and lactation on plasma triglyceride and lipoprotein lipase of adipose and mammary tissue in rats. Parturition occurred on the 22nd or 23rd day of pregnancy. All rats were suckled after parturition. One unit of lipoprotein lipase activity = 1 μmole of chylomicron-triglyceride hydrolyzed to glycerol and FFA per hr. From Hamosh et al. (24).

in the rat on lipoprotein lipase activity in both mammary and adipose tissue and on plasma triglyceride (Table 1). Nonsuckling for 4 hr had no effect, whereas nonsuckling for 9 hr lowered markedly the lipolytic activity in mammary tissue and increased plasma triglyceride concentration from 0.9 to 3.1 mM. Nonsuckling for 18 hr completely inhibited the enzyme activity in mammary tissue and increased the activity in adipose tissue to 55% of that in nonlactating rats; it had no further effect on plasma triglyceride concentration.

TABLE 1. *Effect of nonsuckling and ligation of mammary ducts on lipoprotein lipase activity in the inguinal-abdominal mammary tissue and parametrial adipose tissue and on plasma triglyceride concentration in lactating rats*

| Group | No. of rats | Weight of mammary tissue (unilateral) (g) | Lipoprotein lipase activity | | Plasma triglyceride (mM) |
			Mammary tissue (units/g)	Adipose tissue (units/g)	
Lactating 5–6 days					
Suckled	5	2.8	49 ± 6	1.6 ± 0.2	0.9 ± 0.1
Unsuckled 4 hr	2	4.0	42 (39, 45)	1.6 (1.2, 2.0)	0.6 (0.5, 0.7)
Unsuckled 9 hr	2	4.3	15 (8, 22)	3.4 (3.0, 3.8)	3.1 (2.2, 4.0)
Unsuckled 13 hr	2	4.5	6 (6, 6)	4.2 (2.6, 5.8)	2.5 (2.3, 2.7)
Unsuckled 18 hr	5	4.3	2 ± 0.6	6.5 ± 1.3	3.3 ± 0.9
Ducts ligated 18 hr[a]		6.0	4 ± 1		
	4			1.5 ± 0.3	0.7 ± 0.1
Suckled		2.8	38 ± 8		
Nonlactating	6	0.8	1 ± 0.4	11.8 ± 0.5	0.9 ± 0.04

[a] The ducts draining the inguinal-abdominal glands on one side were ligated for 18 hr while the glands on the other side as well as the pectoral glands were suckled. Values are means ± S.E. From Hamosh et al. (24).

Mammary tissue of lactating animals becomes engorged with milk when the glands are not suckled (Table 1). To determine the effect of local engorgement on lipoprotein lipase activity in mammary tissue, the ducts draining the inguinal-abdominal glands on one side were ligated and the contralacteral glands were left intact, and both groups were assayed 18 hr later for lipoprotein lipase activity (Table 1). Lipoprotein lipase activity was greatly reduced in the duct-ligated glands, to 4 units/g, and the weight was doubled, whereas the enzyme activity and weight were unchanged in the suckled intact glands. The duct ligation had no effect on either plasma triglyceride concentration or lipoprotein lipase activity in adipose tissue.

The localized suppression of lipoprotein lipase activity in the duct-ligated glands with normal levels of activity in the other glands indicates that the effects of ligation are probably independent of circulating hormones. It is possible that the decrease in lipoprotein lipase in the mammary gland at parturition (Fig. 3) was also caused by local engorgement, the result of increased milk production during the last day of pregnancy (25, 26).

The inverse relationship between plasma triglyceride concentration and lipoprotein lipase activity in mammary gland during lactation (Fig. 3 and Table 1) suggests that lipoprotein lipase is involved in the uptake of plasma triglyceride by mammary tissue. The inverse relationship between lipoprotein lipase activity in adipose tissue and that in mammary tissue (Fig. 3 and Table 1) suggests that the hormones responsible for milk secretion may direct dietary fatty acids to mammary tissue by stimulating lipoprotein lipase activity in mammary tissue and suppressing the activity in adipose tissue.

B. Hormonal Control During Lactation

Secretion of milk in the rat is dependent on at least two anterior pituitary hormones, prolactin and ACTH (27, 28). Prolactin acts directly on mammary tissue, whereas ACTH acts indirectly by stimulating the secretion of adrenal glucocorticoid (27). To determine the role of these hormones in the regulation of lipoprotein lipase activity in mammary and adipose tissue during lactation, the effect of the hormones was studied in lactating rats hypophysectomized on the fifth or sixth day of lactation. Since the posterior pituitary gland was also removed, the rats were given oxytocin to ensure milk ejection and thus prevent accumulation of milk in the mammary gland (28, 29). The hormones were injected four times daily for 2 days, starting 2 hr after hypophysectomy. Lipoprotein lipase activity in the tissues (24) was measured 2 hr after the last injection of hormones. The litter sizes were adjusted 2 days before the experiment so that each rat suckled 8 pups.

The effect of hypophysectomy on lipoprotein lipase activity in mammary and adipose tissue of lactating rats is shown in Table 2. Within 6 hr of hypophysectomy the activity in mammary tissue decreased from 50 to 8 units/g. Hypophysectomy also increased, but not until the second day, the activity in adipose tissue from 2 to 8 units/g. Prolactin, 2 mg/day, had no effect for 12 hr and then within 36 hr increased the activity in mammary tissue to 50 units/g. Prolactin also maintained the activity in adipose tissue at a low level, 4.3 units/g. Thus, hypophysectomy changed the lipoprotein lipase activities of mammary and adipose tissue in lactating rats to the levels in nonlactating rats (Table 3), and prolactin injections begun soon after hypophysectomy restored the lipolytic activity in mammary tissue and maintained at a low level the activity in adipose tissue (Table 2).

TABLE 2. *Effect of prolactin (2 mg/day) on lipoprotein lipase activity of mammary and adipose tissue in hypophysectomized lactating rats*

| Hours after hypophysectomy | Lipoprotein lipase activity (units/g) | | | |
| | Mammary tissue | | Adipose tissue | |
	Untreated	Treated	Untreated	Treated
0	50 ± 4 (7)	–	2.4 ± 0.6 (7)	–
6	8 ± 3 (3)	–	1.4 ± 1.0 (3)	–
12	4 ± 1 (4)	6 (6, 6) (2)	2.0 ± 1.5 (4)	1.2 (0.3, 2.0) (2)
24	6 (5, 7) (2)	26 ± 6 (5)	2.9 ± 1.0 (3)	0.7 ± 0.3 (5)
48	4 ± 1 (8)	50 ± 5 (4)	8.4 ± 1.8 (8)	4.3 ± 1.6 (4)

Hypophysectomy was performed on the fifth day postpartum and hormone injections, given every 6 hr, were begun 2 hr later. Oxytocin (1 unit/day) was given to both groups of rats. Values are means ± S.E. Number of rats per group is indicated by the figures in parentheses.

The effects of prolactin on lipoprotein lipase activity in mammary and adipose tissue of hypophysectomized lactating rats were not significantly affected by the simultaneous administration of dexamethasone, growth hormone, and thyroxine, or of growth hormone and thyroxine (Table 3). Administration of dexamethasone alone or in conjunction with growth hormone and thyroxine had a small but significant effect on the enzyme activity

TABLE 3. *Effect of hormones on lipoprotein lipase activity in hypophysectomized lactating rats*

| Group | Hormones given | No. of rats | Lipoprotein lipase activity (units/g) | |
			Mammary tissue	Adipose tissue
Normal				
Nonlactating	none	5	7 ± 2	10.8 ± 0.7
Lactating 7–8 days	none	7	50 ± 4	2.4 ± 0.6
Hypophysectomized				
Lactating	Ox	8	4 ± 1	8.4 ± 1.8
Lactating	Ox, P	4	50 ± 4	4.3 ± 1.6
Lactating	Ox, P, Dex, GH, T	10	70 ± 9	1.3 ± 0.2
Lactating	Ox, P, GH, T	5	40 ± 12	1.7 ± 0.5
Lactating	Ox, Dex	2	27 (25, 29)	3.2 (1.8, 4.6)
Lactating	Ox, Dex, GH, T	7	15 ± 4	4.3 ± 1.2

Rats lactating 5 to 6 days were hypophysectomized and then injected with hormones every 6 hr for 2 days. The hormones and amounts given per day were as follows: oxytocin (Ox), 1 unit; ovine prolactin (P), 2 mg; dexamethasone (Dex), 4 μg; bovine growth hormone (GH), 80 μg; and thyroxine (T), 4 μg. The tissues of hypophysectomized rats were analyzed for lipoprotein lipase activity 48 hr after hypophysectomy.

in mammary gland, and maintained the activity low in adipose tissue. [Growth hormone and thyroxin were tested because they facilitate lactation in other species (28)].

These studies show that the anterior pituitary gland controls lipoprotein lipase activity in mammary and adipose tissue during lactation and that this control is mediated primarily through the secretion of prolactin: in the lactating rat, prolactin stimulates lipoprotein lipase activity in the mammary tissue and suppresses the activity in the adipose tissue. Falconer and Fiddler (30) also observed an increase in lipoprotein lipase activity of mammary glands when prolactin was injected intraductally into pseudopregnant rabbits.

Suckling causes an immediate and continuous release of prolactin and oxytocin to the blood stream (28, 29, 31). Prolactin stimulates and maintains the production of milk by the alveolar cells, and oxytocin stimulates the myoepithelial cells surrounding the alveoli to contract and thus eject milk from the alveolar lumen (28). Accumulation of milk in the gland occurs when oxytocin is lacking (nonsuckling) or when the mammary ducts are ligated. The resultant mechanical compression of capillaries in the tissue undoubtedly reduces delivery of prolactin to the cells and thereby decreases lipoprotein lipase activity in the gland. Lipoprotein lipase activity of mammary tissue is probably present in both alveolar epithelial and capillary endothelial cells (see Section IV). The site of synthesis of the enzyme, however, is not known.

IV. STUDIES IN PERFUSED MAMMARY TISSUES

Studies in lactating guinea pigs (32) and goats (7) infused with chylomicrons containing doubly labeled triglyceride have shown that plasma triglyceride is hydrolyzed during uptake by mammary tissue. The ratio of labeled fatty acids to labeled glycerol in tissue and milk lipid relative to that in chylomicrons indicated that at least one-third of the triglycerides taken up were hydrolyzed to FFA and glycerol.

We have studied the uptake of chylomicron-triglyceride in perfused rat mammary tissue and related the rate of uptake to the level of lipoprotein lipase activity in the tissue. The left inguinal-abdominal mammary glands of the lactating rat were perfused *in situ* by a technique developed recently in our laboratory (33). The anatomy and blood supply of the glands are shown in Fig. 4-A. The skin and underlying mammary glands on the left side have been separated from the abdominal wall and retracted laterally to show the major blood vessels that carry blood to and from the tissue. The mammary tissue was isolated by ligating the following arteries and veins:

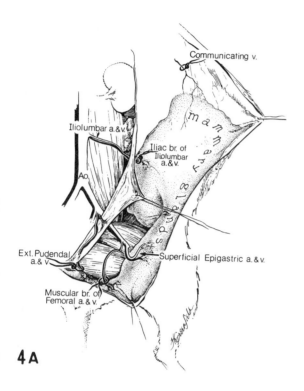

4 A

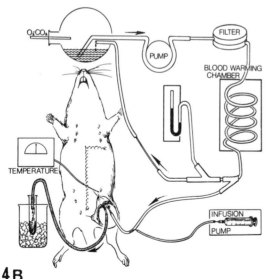

4 B

FIG. 4. A. Anatomy and blood supply of the inguinal-abdominal mammary tissue in the rat. From Mendelson and Scow (33).

B. Schematic diagram of the apparatus used to perfuse rat inguinal-abdominal mammary tissue. The tissue was isolated *in situ* by ligating the muscular branches of the femoral, the external pudendal, and the iliac branches of the iliolumbar artery and vein (see Fig. 4-A). The tissue was perfused through a cannula inserted into the superficial epigastric artery, and venous blood was collected from the tissue through a cannula inserted in the superficial epigastric vein. The venous blood was collected in ice-chilled tubes. From Mendelson and Scow (33).

the muscular branch of the femoral, the external pudendal, and the iliac branches of the iliolumbar. The tissue was perfused through a cannula inserted into the superficial epigastric artery, and venous blood was collected from a cannula inserted into the superficial epigastric vein (Fig. 4-B).

A schematic diagram of the apparatus used to perfuse the inguinal-abdominal mammary tissue of the rat is shown in Fig. 4-B. The apparatus consisted of an oxygenator-reservoir, pump, filter, blood-warming chamber, manometer, and injection site. The bifurcation beyond the warming chamber allowed the blood either to go to the tissue or to return to the reservoir via the bypass tubing. Venous blood was collected continuously in ice-chilled tubes. Substances to be infused were injected intraarterially through a small needle inserted into the arterial tubing. The arterial pressure was maintained at 85 to 95 mm Hg and venous pressure at 1 cm H_2O. The temperature of the tissue, measured with a thermistor probe inserted between the mammary gland and abdominal wall, was maintained at 38°C by warming the blood before it entered the tissue and by warming the tissue with a heat lamp.

The perfusing fluid consisted of a 20% suspension of washed bovine red blood cells in Tyrode's solution containing 4% bovine albumin, 0.1% glucose, and antibiotics (33). Doubly labeled chylomicrons were prepared from chyle collected for 6 hr from the thoracic duct of rats fed corn oil containing [^{14}C]palmitic acid and triolein labeled with [^{3}H]glycerol. The chylomicrons were concentrated by centrifugation at 60,000 g for 1 hr at 5°C, and suspended in 4% albumin solution (34).

The results of a typical perfusion of lactating mammary tissue are shown in Fig. 5. The tissue weighed 4 g, and blood flow averaged 3 ml/min, or 0.75 ml/g per min. Doubly labeled chylomicron-triglyceride was infused at a rate of 0.6 μmoles/min for the first 3 min, 1.5 μmoles/min for the next 2 min, and 3.2 μmoles/min for the last 8 min. Labeled glyceride, FFA, and glycerol were measured in blood samples collected throughout the perfusion.

The release of FFA and glycerol to venous blood by the tissue began during the first 2 min of infusion of chylomicrons and increased as the arterial triglyceride concentration increased (Fig. 5). When the infusion was stopped, the venous triglyceride concentration decreased 98% and the release of FFA decreased 60% in 3 min, whereas the release of glycerol decreased at a much slower rate, 7% in 3 min. The molar ratio of FFA to glycerol in the venous blood was 14.6 during the first 2 min of infusion of chylomicrons, about 5 during the next 11 min, and less than 2 after the infusion was stopped. The very high ratio of FFA to glycerol in venous blood during the first 2 min of infusion suggests that partial glycerides were formed during the initial step(s) of uptake. Partial glycerides, however, were not

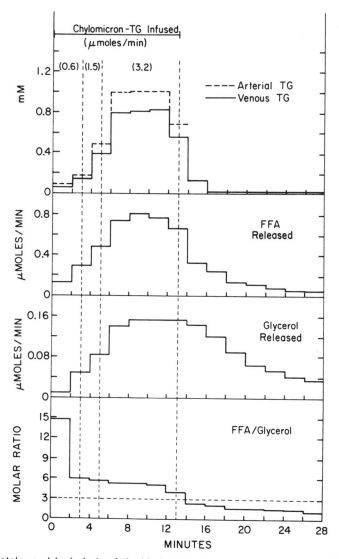

FIG. 5. Uptake and hydrolysis of doubly labeled chylomicron-triglyceride (TG) by per-
fused lactating rat mammary tissue. Venous blood was analyzed for labeled glyceride,
FFA, and glycerol content. The tissue weighed 4 g and blood flow averaged 3 ml/min.
From Mendelson and Scow (33).

TABLE 4. *Effect of nonsuckling on blood flow and metabolism of chylomicron-triglyceride in perfused mammary tissue of lactating rats*[a]

| | | | | Chylomicron-triglyceride | | | |
| | | | | Fatty acids recovered as | | Glycerol recovered as | |
Group	Blood flow (ml/min)	Amount infused (μmoles)	Arterial blood conc[b] (mM)	Plasma FFA (% of infused)	Tissue glyceride (% of infused)	Plasma glycerol (% of infused)	Tissue glyceride (% of infused)
Suckled	2.8 ± 0.2	27 ± 1	0.72	7.6 ± 1.4	10.3 ± 2.5	6.0 ± 1.2	9.6 ± 1.9
Unsuckled 18 hr[c]	1.8 ± 0.2	10 ± 2	0.36	1.5 ± 0.4	1.8 ± 0.3	0.9 ± 0.3	1.7 ± 0.4

[a] Inguinal-abdominal mammary tissue of rats lactating for 5 to 10 days was perfused 15 min with blood, 16 min with blood containing chylomicrons, and then 15 min with blood only. Chylomicron-triglyceride was labeled with [¹⁴C]palmitate and [³H]glycerol. Venous blood samples were collected at 2-min intervals throughout the perfusion. Values are means ± S.E. of four experiments. From Mendelson and Scow (33).

[b] Arterial blood triglyceride concentration during infusion of chylomicrons.

[c] Unsuckled rats were separated from their litters 18 hr prior to perfusion.

released to the blood. The continued release of glycerol after the infusion was stopped indicates that hydrolysis occurs outside of the bloodstream.

The perfused mammary tissue of lactating rats removed from the bloodstream 16% of the chylomicron-triglyceride infused (Table 4). About 40% of the triglyceride taken up was hydrolyzed and released to the blood as FFA and glycerol, and the rest was retained in the tissue.

Uptake of chylomicron-triglyceride was also studied in perfused mammary tissue of rats unsuckled for 18 hr, because nonsuckling for this length of time suppresses almost completely the lipoprotein lipase activity in the tissue (Table 1). Triglyceride uptake, release of FFA and glycerol to the blood, and retention of fatty acids and glycerol were reduced 80% in perfused mammary tissue of unsuckled rats (Table 4). Blood flow through the tissues was also decreased. These findings suggest that lipoprotein lipase activity is involved in the uptake of triglyceride by mammary tissue.

Lipoprotein lipase activity has been found in trace amounts in venous blood of mammary tissue in lactating goats (17) and in capillaries of mammary tissue in lactating rodents (35). Lipoprotein lipase activity is also found in the milk of several species (19, 36, 37), indicating that the enzyme is present in the secretory cells of mammary tissue. The venous blood collected from perfused rat mammary tissue contained a small amount of lipolytic activity, about 0.02 units/ml (Table 5). Infusion of heparin, at a blood concentration of 30 to 40 μg/ml, increased at once the release of lipoprotein lipase activity from 0.02 to 1.5 units/30 sec (Table 5). The amount of ac-

TABLE 5. *Effect of heparin on the release of lipoprotein lipase activity by perfused lactating rat mammary tissue*

Time (min)	Heparin infused[a]	Lipoprotein lipase activity released to the blood stream per 30 sec[b] (Units)
0–5	−	0.02 (0.01, 0.03)
5.0–5.5	+	1.52 (1.50, 1.54)
5.5–6.0	+	2.25 (2.14, 2.36)
6.0–6.5	+	1.84 (1.52, 2.16)
6.5–7.0	+	1.68 (1.50, 1.86)
7.0–7.5	+	1.67 (1.49, 1.85)
7.5–8.0	+	1.24 (1.10, 1.38)

[a] Heparin concentration in the blood was 30 to 40 μg/ml during the infusion of heparin.

[b] One unit of lipoprotein lipase activity = 1 μmole of chylomicron-triglyceride hydrolyzed to glycerol and FFA per hr. Values are means of two experiments.

tivity released during the first 3 min of heparin infusion was equivalent to about 5% of the total activity present in the gland. The large and rapid release of enzyme activity to the blood when heparin was infused indicates that lipoprotein lipase is also present, in appreciable amounts, in the capillary endothelial cells of mammary tissue.

The ratio of labeled fatty acids to labeled glycerol retained by perfused rat mammary tissue was nearly the same as that in chylomicrons (Table 4). This could result from the incorporation into tissue lipid of FFA and partial glycerides, or glycerol, produced by hydrolysis, or from the retention of unhydrolyzed chylomicrons in the vascular bed of the tissue. The latter seems unlikely, however, because the tissue was perfused with chylomicron-free blood for 15 min after the infusion of chylomicrons was stopped, and the amounts of labeled fatty acids and glycerol retained were related to the amounts released as FFA and glycerol (Table 4). Glycerokinase has been found in mammary tissue of the guinea pig (38), suggesting that glycerol, derived from the hydrolysis of chylomicron-triglyceride, could be utilized by mammary tissue for the synthesis of triglyceride. Studies in the goat, however, indicate that incorporation of glycerol into milk triglyceride is negligible (7, 39). In perfused rat mammary tissue, the continued release of glycerol and the decreased ratio of FFA to glycerol in venous blood after the infusion of chylomicrons was stopped (Fig. 5) also suggest poor utilization of glycerol by mammary tissue. Studies in the goat and cow suggest that blood triglyceride is hydrolyzed by mammary tissue to 2-monoglyceride and FFA, and that the monoglyceride is used for the synthesis of milk triglyceride (40, 41). The finding in perfused rat mammary tissue that the molar ratio of FFA to glycerol in venous blood was nearly 15 during the first few minutes of infusion of chylomicrons (Fig. 5) indicates that partial glycerides are formed in rat mammary tissue during the uptake of triglyceride from blood. Additional studies are needed to establish whether 2-monoglyceride, or diglyceride, can be utilized by rat mammary tissue for the synthesis of milk lipid.

FIG. 6. A. Electron micrograph of lactating rat mammary gland showing a mammary alveolar cell (MA) secreting a milk lipid droplet (ML), a myoepithelial cell (Me), and a capillary (E) with endothelial processes (P) projecting into the capillary lumen (L). ×12,000. From E. J. Blanchette-Mackie and R. O. Scow, unpublished.

B. Electron micrograph of mammary tissue of a lactating rat taken 10 min after the intravenous injection of rat chylomicrons. Chylomicrons (C) are seen in the capillary lumen enmeshed by finger-like projections (P) of the capillary endothelium (E). Red blood cell (RBC), myoepithelial cell (Me), mammary alveolar cell (MA), and basement membrane (BM). ×9000. From E. J. Blanchette-Mackie and R. O. Scow, unpublished.

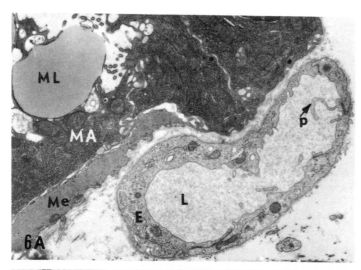

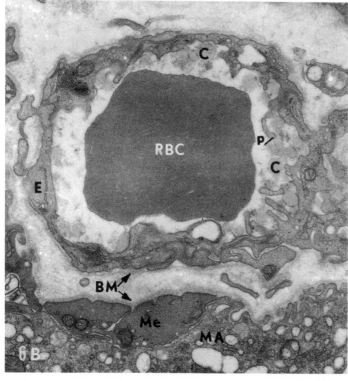

V. TRANSPORT OF CHYLOMICRON-FATTY ACIDS
ACROSS THE MAMMARY CAPILLARY ENDOTHELIUM

The capillaries in mammary tissue consist of a continuous endothelium surrounded by a continuous basement membrane (Figs. 6-A and 6-B) and pericytes (Fig. 7-C). The endothelial cells contain numerous microvesicles, 0.06 to 0.08 μ in diameter, and vacuoles, up to 0.25 μ in diameter (Figs. 7-B and 7-C). The endothelial cells also have processes, located along the luminal surface, which project into the capillary lumen (Figs. 6-A, 6-B, and 7-A). Chylomicrons were found attached to the luminal surface (Figs. 6-B, 7-A, and 7-B), enmeshed by the endothelial projections (Figs. 6-B and 7-A), and partially enveloped by the capillary endothelium (Fig. 8-B), in lactating animals injected intravenously with chylomicrons (35). Some of the enveloped chylomicrons were seen in direct contact with microvesicles (Fig. 8-B). No material resembling fat was seen within microvesicles or vacuoles, in the junctions between endothelial cells, or in the space outside of the capillaries (35).

Schoefl and French (35), using a cytochemical technique, have demonstrated lipoprotein lipase activity in the capillaries of lactating mammary tissue. The activity was associated with chylomicrons attached to the luminal surface of the endothelium. A similar technique was used in our laboratory to locate sites of lipoprotein lipase activity in adipose tissue (42). It was found that glyceride derived from chylomicrons was hydrolyzed to FFA in the capillary endothelial cells and in the subendothelial space between the endothelium and pericytes. The chemical nature of the glyceride in the endothelium and in the subendothelial space, however, was not determined. On the basis of both cytochemical (42) and biochemical (15) findings in perfused adipose tissue, it was proposed that chylomicron-triglyceride is hydrolyzed to diglyceride and FFA at the luminal surface of the endothelium, that diglyceride is hydrolyzed to monoglyceride and FFA

FIG. 7. A. Electron micrograph of lactating rat mammary gland taken 10 min after the intravenous injection of chylomicrons. Note chylomicrons (C) along the capillary endothelium (E), some of which are enveloped by a finger-like projection of the endothelium (P). Basement membrane (BM), mammary alveolar cell (MA), and capillary lumen (L). ×20,000. From E. J. Blanchette-Mackie and R. O. Scow, unpublished.

B. Same as Fig. 7-A. Numerous chylomicrons (C) are present along the luminal surface of the endothelium (E). Microvesicles (mV), basement membrane (BM), and mammary alveolar cell (MA). ×20,000. From E. J. Blanchette-Mackie and R. O. Scow, unpublished.

C. Same as Fig. 7-A. Note numerous vacuoles (V) in the endothelium (E). Lumen (L), pericyte (Pe), and basement membrane (BM) ×25,000. From E. J. Blanchette-Mackie and R. O. Scow, unpublished.

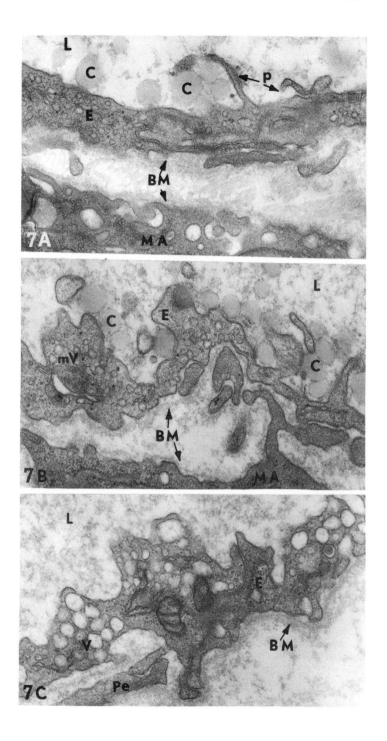

in vacuoles and microvesicles, and that monoglyceride is hydrolyzed to glycerol and FFA in the subendothelial space (15).

Redgrave (43) reported that lipolysis of chylomicrons *in vivo* results in smaller particles which contain the same amount of cholesterol ester but less triglyceride. The structure of the particles was not described. It has been suggested that these "remnants" (43), or "skeletons" (44), are removed from the blood by the liver (43, 44).

Triglyceride-depleted chylomicrons have been produced *in vitro* by incubating chylomicrons with lipoprotein lipase (45). Electron microscopic studies showed that loss of triglyceride during lipolysis resulted in collapsed spheres consisting of the initial surface coat and residual core material. Similarly collapsed spheres have also been found in venous blood plasma collected from lactating rat mammary tissue perfused 2 to 3 min with chylomicrons (Fig. 8-A). These particles, produced by lipolysis within the mammary capillary bed, are undoubtedly chylomicron "remnants" (43).

It is evident from studies in several species that blood triglyceride is hydrolyzed during uptake by mammary tissue and that the glyceryl moiety as well as the fatty acids are incorporated into milk triglyceride (7, 32) (Table 5). Although glycerokinase is present in mammary tissue (38), it seems likely, as discussed above (see Section IV), that 2-monoglyceride, derived by hydrolysis from blood triglyceride, is the primary form in which the glyceryl moiety is taken up by mammary alveolar cells. This would involve transfer of partial glycerides across the extracellular space in mammary tissue, from the capillary endothelium to the alveolar cells.

A schema showing how chylomicron-fatty acids may cross from the mammary capillary lumen to the alveolar cells is shown in Fig. 8-C. It is based primarily on the biochemical and morphological findings in mammary tissue described above and, in part, on our studies in perfused rat adipose tissue (15, 33). The schema depicts what we think is the most likely fate of chylomicron-triglyceride in the capillaries of rat mammary tissue. When a chylomicron becomes attached to and is partially enveloped by the endothelial cell, some of the triglyceride molecules, depicted by three dots in a lined area, are hydrolyzed by lipoprotein lipase to FFA and diglyceride. The FFA

FIG. 8. A. Chylomicrons (C) and chylomicron remnants (CR) in venous blood collected from lactating rat mammary glands perfused with blood containing chylomicrons. Stained with sodium phosphotungstate. ×65,000. From E. J. Blanchette-Mackie, Carole R. Mendelson, and R. O. Scow, unpublished.

B. Electron micrograph of a lactating rat mammary gland showing the relationship of a partially enveloped chylomicron (C) to microvesicles (mV) within the capillary endothelial cell (E). Connective tissue (Cf). ×90,000. From G. I. Schoefl and J. E. French (35).

C. Schema showing mode of transport of chylomicron-fatty acids across the capillary wall in lactating mammary tissue. See text for explanation of symbols.

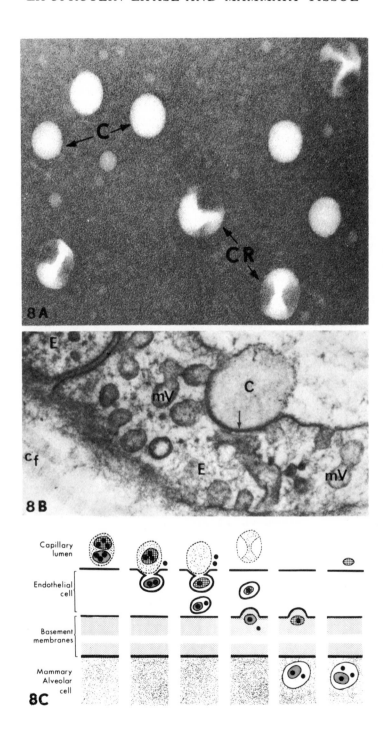

(a single dot) is released to the blood, and the diglyceride (two dots in a lined area) is taken into a microvesicle or vacuole. The diglyceride is then hydrolyzed and the products, FFA and monoglyceride (a single dot in a lined area), are released to the subendothelial space (basement membrane area), to be taken up and esterified by the mammary alveolar cell. Other chylomicron-triglyceride molecules, depicted by three dots in a crosshatched area, are hydrolyzed at the luminal surface to FFA and monoglyceride. The FFA are released to the blood, and the monoglyceride (a single dot in a crosshatched area) is transported across the endothelial cell, within a microvesicle or vacuole, to the subendothelial space where it is hydrolyzed to FFA and glycerol. The FFA is then taken up by the mammary alveolar cell to be esterified, and the glycerol (the crosshatched area) is released to the bloodstream. The molar ratio of FFA to glycerol released to the blood is 3, and the ratio of fatty acids to glycerol taken up by the alveolar cells is also 3 (Table 5). The chylomicron "remnant" (depicted by the stippled hourglass-shaped particle), consisting of the surface coat and the residual triglyceride core, is released to the blood and is subsequently taken up by the liver. Some of the FFA and glycerol released may be incorporated by the liver into very low density lipoprotein and then taken up by the mammary tissue. Uptake of very low density lipoprotein-triglyceride also involves lipoprotein lipase (10, 11).

VI. CONCLUSIONS

Most of the long-chain fatty acids secreted in milk are of dietary origin and are transported to the mammary tissue in the bloodstream as triglyceride. Triglyceride in blood is also taken up by other tissues, such as muscle and adipose tissue. There is considerable evidence that during uptake blood triglyceride is hydrolyzed in the capillary wall to FFA and glycerol by lipoprotein lipase.

Lipoprotein lipase activity in mammary tissue is low in nonpregnant rats, increases during late pregnancy, and remains high throughout lactation. In contrast, lipoprotein lipase activity in adipose tissue is high in nonpregnant rats and during most of pregnancy, decreases 2 to 3 days before parturition, and remains low during lactation. Nonsuckling for 18 hr decreases markedly lipoprotein lipase activity in mammary tissue and restores the activity in adipose tissue. Studies in hypophysectomized lactating rats given different hormones indicate that the changes in lipoprotein lipase activity in mammary and adipose tissue during lactation are mediated through the secretion of prolactin by the pituitary gland. These changes serve to direct dietary fatty acids to mammary tissue for the formation of milk lipid.

Studies in perfused mammary tissue of suckled and unsuckled rats show that chylomicron-triglyceride is hydrolyzed during uptake and that the amount taken up is related to the level of lipoprotein lipase activity in the tissue. The findings also show that partial glycerides are formed during the initial step(s) of uptake, and that the glyceryl moiety as well as the fatty acids are retained by the tissue. It is proposed that partial glycerides and FFA, formed by the hydrolysis of chylomicron-triglyceride, are transported across the capillary endothelium and extracellular space to the alveolar cells for the synthesis of milk.

ACKNOWLEDGMENT

The authors are grateful to Mr. Barker Dixon, Mrs. Theresa R. Clary Fleck, Mr. David Winter, and Mrs. Gloria Devore for their expert assistance in these studies.

REFERENCES

1. Garton, G. A. (1963): *J. Lipid Res.*, 4:237.
2. Breckenridge, W. C., and Kuksis, A. (1967): *J. Lipid Res.*, 8:473.
3. Insull, W., Jr., Hirsch, J., James, T., and Ahrens, E. H., Jr. (1959): *J. Clin. Invest.*, 38:443.
4. Beare, J. L., Gregory, E. R. W., Smith, D. M., and Campbell, J. A. (1961): *Can. J. Biochem. Physiol.*, 39:195.
5. Carey, E. M., and Kils, R. (1972): *Biochem. J.*, 126:1005.
6. Annison, E. F., Linzell, J. L., Fazakerley, S., and Nichols, B. W. (1967): *Biochem. J.*, 102:637.
7. West, C. E., Bickerstaffe, R., Annison, E. F., and Linzell, J. L. (1972): *Biochem. J.*, 126:477.
8. Glascock, R. F., Welch, V. A., Bishop, C., Davies, T., Wright, E. W., and Noble, R. C. (1966): *Biochem. J.*, 98:149.
9. Bishop, C., Davies, T., Glascock, R. F., and Welch, V. A. (1969): *Biochem. J.*, 113:629.
10. Fredrickson, D. S., Levy, R. I., and Lees, R. S. (1967): *New Eng. J. Med.*, 276:34, 94.
11. Robinson, D. S. (1970): *Compr. Biochem.*, 18:51.
12. Bragdon, J. H., and Gordon, R. S., Jr. (1958): *J. Clin. Invest.*, 37:574.
13. Olivecrona, T. (1962): *J. Lipid Res.*, 3:1.
14. Olivecrona, T., and Belfrage, P. (1965): *Biochim. Biophys. Acta*, 98:81.
15. Scow, R. O., Hamosh, M., Blanchette-Mackie, E. J., and Evans, A. J. (1972): *Lipids*, 7:497.
16. Bezman, A., Felts, J. M., and Havel, R. (1962): *J. Lipid Res.*, 3:427.
17. Barry, J. M., Bartley, W., Linzell, J. L., and Robinson, D. S. (1963): *Biochem. J.*, 89:6.
18. Brody, S. (1945): In: *Bioenergetics and Growth.* Reinhold, New York, p. 382.
19. McBride, O. W. and Korn, E. D. (1963): *J. Lipid Res.*, 4:17.
20. Robinson, D. S. (1963): *J. Lipid Res.*, 4:21.
21. Ota, K., and Yokoyama, A. (1967): *J. Endocrinol.*, 38:35.
22. Long, J. F. (1969): *Am. J. Physiol.*, 217:228.
23. Otway, S., and Robinson, D. S. (1968): *Biochem. J.*, 106:677.
24. Hamosh, M., Clary, T. R., Chernick, S. S., and Scow, R. O. (1970): *Biochim. Biophys. Acta*, 210:473.

25. Jeffers, K. R. (1935): *Am. J. Anat.*, 56:257.
26. Kuhn, N. J., and Lowenstein, J. M. (1967): *Biochem. J.*, 105:995.
27. Lyons, W. R., Li, C. H., and Johnson, R. E. (1958): *Recent Progr. Hormone Res.*, 14:219.
28. Cowie, A. T. (1961): In: *Milk: The Mammary Gland and Its Secretion*, Vol I, edited by S. K. Kon and A. T. Cowie. Academic Press, New York, p. 163.
29. Cross, B. A. (1961): In: *Milk: The Mammary Gland and Its Secretion*, Vol. I, edited by S. K. Kon and A. T. Cowie. Academic Press, New York, p. 229.
30. Falconer, I. R., and Fiddler, T. J. (1970): *Biochim. Biophys. Acta*, 218:508.
31. Terkel, J., Blake, C. A., and Sawyer, C. H. (1972): *Endocrinology*, 91:49.
32. McBride, O. W., and Korn, E. D. (1964): *J. Lipid Res.*, 5:459.
33. Mendelson, C. L., and Scow, R. O. (1972): *Am. J. Physiol.*, 223:1418.
34. Scow, R. O., Stein, Y., and Stein, O. (1967): *J. Biol. Chem.*, 242:4919.
35. Schoefl, G. I., and French, J. E. (1968): *Proc. Roy. Soc. B.*, 169:153.
36. Korn, E. D. (1962): *J. Lipid Res.*, 3:246.
37. Hamosh, M., and Scow, R. O. (1971): *Biochim. Biophys. Acta*, 231:283.
38. McBride, O. W., and Korn, E. D. (1964): *J. Lipid Res.*, 5:442.
39. Pynadath, T. I., and Kumar, S. (1964): *Biochim. Biophys. Acta*, 84:251.
40. Dimick, P. S., McCarthy, R. D., and Patton, S. (1965): *Biochim. Biophys. Acta*, 116:159.
41. Bickerstaffe, R., Linzell, J. L., Morris, L. J., and Annison, E. F. (1970): *Biochem. J.*, 117:39P.
42. Blanchette-Mackie, E. J., and Scow, R. O. (1971): *J. Cell Biol.*, 51:1.
43. Redgrave, T. G. (1970): *J. Clin. Invest.*, 49:465.
44. Bergman, E. N., Havel, R. J., Wolfe, B. M., and Bøhmer, T. (1971): *J. Clin. Invest.*, 50: 1831.
45. Scow, R. O., Blanchette-Mackie, E. J., Hamosh, M., and Evans, A. J. (*in* press): *Wiss. Veröffentl. Deut. Ges. Ernährung*, 23.

Dietary Lipids and Postnatal Development
Raven Press, New York © 1973

Mechanism of Fatty Liver in Infantile Malnutrition

Hernando Flores, Anne Seakins*, and Fernando Mönckeberg**

*Departamento de Ciencias de los Alimentos y Tecnología Química, Facultad de Química y Farmacia, Universidad de Chile, Casilla 5370, Santiago 3, Chile, *Tropical Metabolism Research Unit, University of The West Indies, Mona, Kingston 7, Jamaica, and **Laboratorio de Investigaciones Pediátricas, Facultad de Medicina, Universidad de Chile, Casilla 5370, Santiago 3, Chile*

It has been established that there are two extreme types of infantile malnutrition, marasmus and kwashiorkor. The etiology of these conditions is different: marasmus is the result of prolonged survival on diets which are deficient both in calories and protein, whereas in kwashiorkor there is a predominant lack of protein on a relatively normal caloric intake. One of the most characteristic features of kwashiorkor is the enlargement of the liver due to fatty infiltration, and this is commonly used to differentiate the two types of malnutrition.

Up to 50% of the weight of the liver may be fat in kwashiorkor (1). Although liver lipids have seldom been examined, there is universal agreement that the main lipid in the excess liver fat is triglyceride (2). We have recently determined triglycerides directly in liver biopsy specimens from children with kwashiorkor and found that in fact triglycerides account for practically all the liver lipids (Table 1).

TABLE 1. *Comparison between the concentration of total fat and triglycerides in liver biopsies from five Jamaican children with kwashiorkor*

Case	Total fat (g/100 g liver)	Triglycerides
1	41.9	34.0
2	26.8	30.8
3	35.2	19.6
4	34.9	27.1
5	48.7	50.6

115

Several hypotheses have been suggested to explain the pathogenesis of the fatty infiltration of the liver. Observations by different workers on lipid metabolism in this group of malnourished children have agreed remarkably well, although the conclusions drawn from the data have been different.

A deficiency of essential fatty acids as contributory to the accumulation of lipids in the liver has been suggested by Bronte-Stewart (3). Schendel and Hansen (4) observed a pattern of fatty acids in serum from patients recovering from kwashiorkor similar to that of essential fatty acid (EFA) deficiency. However, these authors interpreted this to be the result of the great demand produced by increased mobilization of fat rather than a real deficiency. This view is in agreement with a later study (5) in which no significant changes were found in the pattern of essential fatty acids in liver during recovery from kwashiorkor.

The possibility of a specific deficiency of lipotropic factors was first considered by Waterlow (6). He administered choline, methionine, or inositol as dietary supplements to children being treated for kwashiorkor with no beneficial effects. This observation agrees well with the finding of a relatively normal concentration of liver phospholipids, when allowance is made for the excess liver triglycerides (7). Moreover, total serum phospholipids are only slightly reduced in untreated patients and their rise during recovery is of little significance (5, 8–10).

Fletcher (11) found a decrease in the activity of the enzyme glucose-6-phosphatase in liver biopsies from children with kwashiorkor. He suggested that the inability of the liver to secrete glucose could enhance lipogenesis in this organ. Later Alleyne and Scullard (12) were not able to confirm the low glucose-6-phosphatase activity in the liver. Therefore there seems to be no consistent evidence that liver lipogenesis is increased in kwashiorkor.

Lewis et al. (9) believed that the excess liver fat derives from increased mobilization of free fatty acids from the depots. The evidence of increased levels of free fatty acids in serum in kwashiorkor is indeed quite consistent in several reports (9, 11, 13–16). However we have found in Chile that children with uncomplicated kwashiorkor have normal levels of plasma free fatty acids (17). Vomiting and diarrhea are often reported as accompanying features of kwashiorkor and the stress and physiological starvation that they produce could explain the high levels of free fatty acids, which do not need to be associated with the fatty liver. In accordance with this view, Lewis et al. (9) observed that free fatty acids return rapidly to normal levels on feeding starch and sucrose but no protein.

As mentioned above, the main lipid fraction accumulating in the liver in kwashiorkor is triglyceride. Thus the causes of fatty liver can be related to triglyceride metabolism. The best documented alteration of lipid metabolism in kwashiorkor is a very low level of circulating triglycerides (5, 8–10, 17). It is also well documented that as soon as treatment starts there is a striking rise in serum lipids, especially in triglycerices (5, 8–10, 17–20). This rise during treatment led Schwartz and Dean (8) to postulate that nutritional deficiencies in kwashiorkor could cause the lipids to be "locked" in the liver. This is at present the current hypothesis on the mechanism of the fatty infiltration, as most of the data obtained in later work tend to provide support for it. A relative inability of the liver to dispose of triglycerides has been suggested as a contributory mechanism in almost all the hypotheses offered so far (4, 9, 11). Waterlow (6) has identified protein as the dietary factor most likely to be responsible for the clearing of the liver fat during recovery. Lewis et al. (9) showed that the changes in serum triglycerides did not occur until protein was introduced in the diet.

The main pathway of liver triglycerides is their secretion into plasma in very low-density lipoprotein. It has been shown that low-density lipoproteins are in fact responsible both for the low initial levels of serum triglycerides and for their rise during recovery (10, 17). The hypothesis is, therefore, that the rise in the fasting levels of low-density lipoprotein triglycerides during recovery indicates defatting of the liver. We have measured liver lipids before and after the rise of serum triglycerides to a peak during treatment of children with kwashiorkor and have shown that in effect, liver lipids decrease to approximately 50% of the initial level in this period (17) (Fig. 1).

In experiments with perfused rat livers, Heimberg et al. (21) observed that the components of the d < 1.019 lipoprotein were released into the perfusate in constant proportions. On this basis the authors proposed that the synthesis of very low-density lipoprotein can be inhibited, leading to fatty liver, by blocking the synthesis of any of the components of the lipoprotein.

There are few studies on the composition of the very low-density lipoprotein in man. It has been observed that during recovery from kwashiorkor there is a constant proportion of triglycerides, phospholipids, and cholesterol in the d < 1.063 serum lipoprotein fraction (17). On the other hand, Schonfeld (22) reported that during carbohydrate induction in hyperlipemic patients there is a rise in the triglyceride:protein ratio of very low-density lipoprotein, which indicates that the lipoprotein can have an abnormal triglyceride load. We have examined the composition of the very low-density lipoprotein of children recovering from kwashiorkor, as well as of recovered children and normal adults under different dietary conditions. When the concentrations of triglycerides are compared with those of protein, phospholipids, and cholesterol, highly significant correlation coefficients are obtained (Fig. 2). Moreover, the data obtained by Schonfeld (22) from normal individuals seem to fit remarkably well in our regression equation.

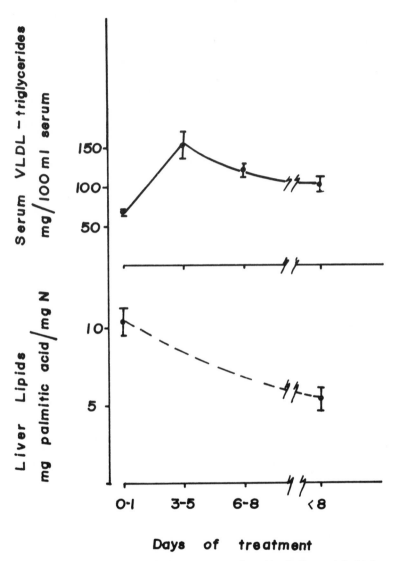

FIG. 1. Relationship between changes in serum very low-density lipoprotein triglycerides and liver lipids during recovery from kwashiorkor. Adapted from Flores et al. (17). By courtesy of the Editor, *Br. J. Nutr.*

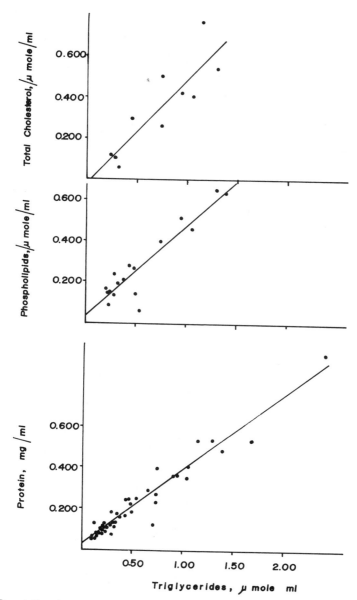

FIG. 2. Correlation between the concentrations of the different components of human serum very low-density lipoprotein. The simple correlation coefficients were: Triglyceride: protein, 0.97 (y = 0.36 x + 0.0315); triglyceride:phospholipid, 0.92 (y = 0.44 x + 0.0373); triglyceride:cholesterol, 0.87 (y = 0.50 x − 0.0126). The subjects were children recovered from malnutrition and normal adults fed diets with different fat contents between 20% and 70% of the total calories, as well as children recovering from kwashiorkor.

Thus it appears that with the exception of hyperlipemic patients, the very low-density lipoprotein behaves as a definite chemical entity, and the suggestion of Heimberg et al. (21) would also be applicable to humans. This evidence validates the hypothesis that in kwashiorkor the impairment in the synthesis of very low-density lipoprotein lies at the level of lipoprotein apopeptide synthesis (17). This hypothesis was based on the known low-protein intake of kwashiorkor and on the association found between the presence of fatty liver and very low levels of serum B-globulin in malnourished children (17). This globulin fraction probably contains the circulating apoprotein of very low-density lipoprotein (23) described by Roheim et al. (24).

Obvious limitations pose serious difficulties to obtaining further evidence on the hypothesis reviewed above for human subjects. Therefore, we decided to study the possibility of obtaining a model of kwashiorkor in the rat suitable for experimentation. Such a model was obtained by feeding a protein-free diet to weanling rats for 15 days (25). The animals developed fatty livers and showed most of the characteristics of human kwashiorkor, including the low levels of fasting serum triglycerides. We injected these animals with the serum protein fraction which contains the very low-density

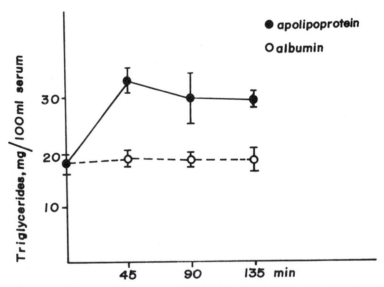

FIG. 3. Effect of apolipoprotein or albumin on the levels of fasting serum triglycerides of protein-depleted rats with fatty livers. From Flores et al. (25). By courtesy of the Editor, *J. Nutr.*

lipoprotein apopeptide described by Roheim et al. (24). This treatment produced a rapid increase in the levels of fasting serum triglycerides (Fig. 3) and markedly enhanced the incorporation of labeled oleic acid into serum triglycerides (Table 2). These effects were not observed in normal animals, and it was concluded that in the protein-depleted rats the protein precursor of very low-density lipoprotein is limiting for its synthesis.

TABLE 2. *Effect of apolipoprotein on the incorporation of ^{131}I-oleic acid into plasma triglycerides*

	^{131}I-oleic acid incorporation, cpm/μmole triglyceride	
Treatment	Normal rats	Protein-depleted rats
apolipoprotein	18,176 ± 2,873	35,417 ± 4,665
albumin	29,982 ± 7,900	9,335 ± 2,000
p	0.10	0.05

From Flores et al. (27). By courtesy of the Editor, *J. Nutr.* ^{131}I-oleic acid was injected intramuscularly at time 0. At 60 min the rats were injected the plasma protein fraction containing the apolipoprotein or isologous albumin. All animals were killed at 150 min.

In later experiments we observed that some rats fail to develop fatty liver when fed the protein-free diet, although no differences in the dietary intake could be demonstrated. A significant correlation was found between the body weight of the rats at weaning and their ability to develop fatty liver (Fig. 4). The reason for the differences of response of the underweight and normal rats to protein depletion remains to be elucidated, but, as regards the absence of fatty liver, a preliminary approach has been the measurement of liver triglyceride synthesis. The incorporation of a labeled fatty acid into liver triglycerides was markedly decreased in the underweight rats fed the protein-free diet, as compared to normal controls (Table 3). We believe that this can account for the absence of fatty liver as the decreased synthesis of triglycerides in the liver can compensate for the decreased synthesis of very low-density lipoprotein. Thus it is apparent that the nutritional status is an important variable in determining the alterations of lipid metabolism produced by protein depletion.

Although the evidence in favor of decreased synthesis of the protein moiety of very low-density lipoprotein as the primary pathogenic mechanism of fatty liver in kwashiorkor is abundant and convincing, patterns of lipid metabolism different from that presented above also exist. Coward and Whitehead (26) have found some patients in Uganda with kwashiorkor who had high levels of serum B-lipoprotein. In Jamaica we have found that three types of patients can be distinguished in this respect, having low, normal, and high levels of circulating triglycerides. Although at present no explanation exists for this unusual finding, it indicates that factors other than decreased protein synthesis alone can modify lipid metabolism in infantile

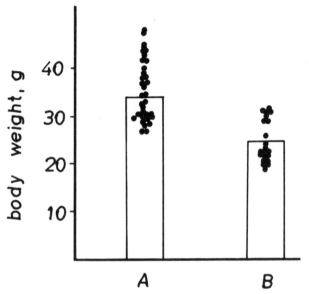

FIG. 4. Relationship between body weight of rats at weanling and ability to develop fatty liver on a protein-free diet. Group A: rats which developed fatty livers; Group B: underweight rats which failed to develop fatty liver.

TABLE 3. [131]*I-oleic acid incorporation into liver triglycerides in rats*

	Normal	Underweight	Underweight protein-depleted
Liver triglycerides (mg/g)	5.1	2.1	1.4
[131]I-oleic acid incorporation into triglycerides, % of the dose	2.4	0.7	0.49

Normal and underweight rats were selected as shown in Fig. 4. The animals were killed 10 min after the intravenous injection of albumin-bound ^{131}I-oleic acid. Liver triglycerides were isolated by thin-layer chromatography.

malnutrition. In our patients with normal and low levels of fasting serum triglycerides a significant relationship exists between these levels and the age of the patients (Fig. 5). Differences in the degree of iron depletion of the children could also contribute to altering the pattern of lipid metabolism: Amine and Hegsted (27) have found that iron deficiency produces marked hypertriglyceridemia in rats and chicks.

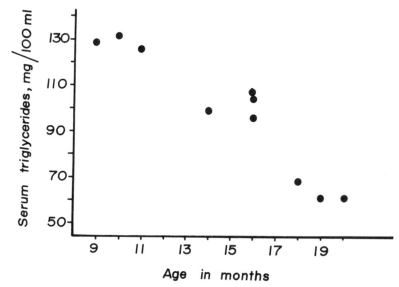

FIG. 5. Relationship between the levels of fasting serum triglycerides and the age of Jamaican children with kwashiorkor. Fasting serum triglycerides were determined on the day following admission of the patients to hospital. A group of four patients who presented values over 200 mg/100 ml serum did not fit into this correlation.

An important factor in the control of the level of fasting serum triglycerides is the amount of dietary lipids. Children fed a diet containing 60% of their calories as fat have very low levels of serum triglycerides, but on change of this diet for one containing 26% of their calories as fat show a rapid increase of the values which stabilize in "normal" values thereafter (Fig. 6). The opposite effect is even more rapid. This phenomenon could be the opposite of the well-known carbohydrate induction. The ease of the response to changes in the dietary fat intake in the range of 26% to 60% of the total calories suggests that the site of control of very low-density lipoprotein synthesis is more sensitive to fat than to carbohydrate intake. It is unlikely to expect differences in the lipid content of the diets leading to

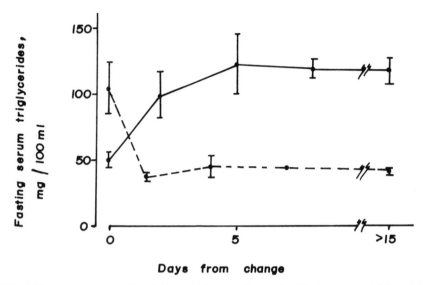

FIG. 6. Effect of changes in the dietary fat intake on the levels of fasting serum triglycerides of children. Solid line: children who were being fed a diet containing 60% of fat calories were changed at day 0 onto a diet containing 26% of fat calories. The dashed line shows the reverse dietary change.

kwashiorkor, but this effect of dietary fat should be considered in the treatment of patients as it may affect the rates of clearing of liver lipids, and when setting normal limits for fasting serum triglycerides.

ACKNOWLEDGMENTS

Part of this work was done in the Tropical Metabolism Research Unit, University of the West Indies, Jamaica, under the tenure of a fellowship from the Wellcome Trust, London, by H. F. H. F. is indebted to the Ford Foundation for a travel grant to attend this Symposium. G. Mutebi collaborated in the measurements of liver triglyceride synthesis.

REFERENCES

1. Waterlow, J. C., and Alleyne, G. A. O. (1971): *Adv. Protein Chem.,* 25:117.
2. MacDonald, I. (1960): *Metabolism,* 9:838.
3. Bronte-Stewart, B. (1961): In: *Recent Advances in Human Nutrition,* edited by J. F. Brock. Churchill, London.
4. Schendel, H. E., and Hansen, J. D. L. (1961): *Am. J. Clin. Nutr.,* 9:735.
5. MacDonald, I., Hansen, J. D. L., and Bronte-Stewart, B. (1963): *Clin. Sci.,* 24:55.
6. Waterlow, J. C. (1948): *Medical Research Council, Special Report,* Series No. 263. London.

7. Chatterjee, K. K., and Mukherjee, K. L. (1968): *Br. J. Nutr.,* 22:145.
8. Schwartz, R., and Dean, R. F. A. (1957): *J. Trop. Pediat.,* 3:23.
9. Lewis, B., Hansen, J. D. L., Wittman, W., Krut, L. H., and Stewart, F. (1964): *Am. J. Clin. Nutr.,* 15:161.
10. Truswell, A. S., Hansen, J. D. L., Watson, C. E., and Wannenburg, P. (1969): *Am. J. Clin. Nutr.,* 22:568.
11. Fletcher, K. (1966): *Am. J. Clin. Nutr.,* 19:170.
12. Alleyne, G. A. O., and Scullard, G. H. (1969): *Clin. Sci.,* 37:631.
13. Lewis, B., Wittman, W., Krut, L. H., Hansen, J. D. L., and Brock, J. F. (1966): *Clin. Sci.,* 30:371.
14. Rao, K. S. J., and Prasad, P. S. K. (1966): *Am. J. Clin. Nutr.,* 19:205.
15. Hadden, D. R. (1967): *Lancet,* 2:589.
16. Milner, R. D. G. (1971): *Pediat. Res.,* 5:33.
17. Flores, H., Pak, N., Maccioni, A., and Mönckeberg, F. (1970): *Br. J. Nutr.* 24:1005.
18. Cravioto, J., De la Pena, C. L., and Burgos, G. (1969): *Metabolism,* 8:722.
19. Mönckeberg, F. (1966): *Nutrición, Bromatología Toxicología,* 5:31.
20. Mönckeberg, F. (1968): In: *Calorie Deficiencies and Protein Deficiencies,* edited by R. A. McCance and E. M. Widdowson. J. and A. Churchill Ltd., London.
21. Heimberg, M., Weinstein, I., Dishmon, G., and Fried, M. (1965): *Am. J. Physiol.,* 209: 1053.
22. Schonfeld, G. (1970): *J. Lab. Clin. Med.,* 75:206.
23. Lees, R. S. (1967): *J. Lipid Res.,* 8:396.
24. Roheim, P. S., Miller, L., and Eder, H. A. (1965): *J. Biol. Chem.,* 240:2994.
25. Flores, H., Sierralta, W., and Mönckeberg, F. (1970): *J. Nutr.,* 100:375.
26. Coward, W. A., and Whitehead, R. G. (1971): *Br. J. Nutr.,* 27:383.
27. Amine, E. K., and Hegsted, D. M. (1971): *J. Nutr.,* 101:1575.

Dietary Lipids and Postnatal Development
Raven Press, New York © 1973

Essential Fatty Acid Deficiency in Humans

Ralph T. Holman

The Hormel Institute, University of Minnesota, Austin, Minnesota 55912

I. INTRODUCTION

The primary purpose of nutritional research on animals is to obtain information and knowledge applicable to man. When a phenomenon is observed in animals, a parallel phenomenon in man is sought. In the case of essential fatty acids (EFA), observation of deficiency in man was delayed so long after its demonstration in animals that a whole generation of physicians sincerely believed that "essential fatty acids," or polyunsaturated acids, played no essential role in human metabolism or nutrition. The vital role of these substances in human metabolism has now been firmly established, and must be considered seriously in the nutrition of humans of all ages. The role of polyunsaturated acids in human nutrition and metabolism has been reviewed recently by the author and his colleagues (1). Thus, a brief historical outline plus description of very recent studies in man will serve to bring the subject to the present.

II. HISTORICAL DEVELOPMENT

Shortly after Burr and Burr (2) described the deficiency state induced by feeding rats a diet free of fat, his laboratory undertook to develop the same deficiency in one man (3). Their negative results, together with earlier, more primitive studies of the effects of a fat-free diet or a low-fat diet in humans, gave a long-lasting impression that man really did not need the essential fatty acids. In hindsight, the negative results are explainable because the diets were not really fat-free or essential fatty acid-free, the durations of the experiments were not long enough to develop a pronounced deficiency in an adult, and the low-grade deficiencies which were actually developed were not recognized.

The first convincing evidence that essential fatty acids were beneficial, or possibly required by humans, came in the publications of Hansen and his co-workers (4–6), who found that supplements of lard, corn oil, or linseed oil to the diets of children suffering with various skin disorders were beneficial to some of the patients. Seventy-five percent of the patients with intractable eczema responded to such supplements, and the eczematous patients were found to have serum lipids less unsaturated than those of a control group. Following supplementation of the diets of the eczematous children, the unsaturation of serum lipids increased and became normal, and the eczema improved.

Meanwhile, animal experimentation had shown that linoleic acid was required for growth and reproduction, and that deprivation of this acid led to a number of malfunctions in animals (7). The list of species which required essential fatty acids grew rather large, and extended from insects to mammals. The primary essential fatty acid was found to be linoleic acid. The metabolic pathways for linoleic acid in animal systems were worked out (Fig. 1), and the existence of several families of polyunsaturated acids which are not interconvertible in higher animals was proven. That is, linoleic acid is the precursor of one family of fatty acids (ω6) and linolenic acid is the precursor of a different family of fatty acids (ω3). Oleic acid, which is formed endogenously from nonfat precursors, is the parent of yet another family of acids (ω9), which increases in the fat-deficient animal because of lack of ω6 and ω3 precursors.

The pattern of polyunsaturated acids in the lipids of an individual is thus strongly influenced by the levels of the various polyunsaturated acids in its diet. If a fat-free diet (devoid of ω6 and ω3 acids) is fed, polyunsaturated acids of these families cannot be synthesized and laid down in tissue lipids. In this condition, the predominant polyunsaturated acids of tissue became

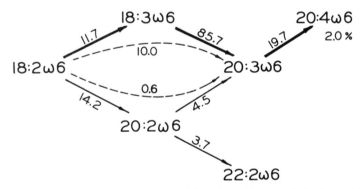

FIG. 1. Metabolic pathways for conversion of linoleic acid to arachidonic acid. Relative yields of reactions were measured *in vitro* under standardized conditions, using microsomes from livers of fat-fed rats and incubated in the presence of all fatty acids found in liver microsomes. The preferred pathway is indicated by the heavy arrows (34).

those of the ω9 group, which can be synthesized endogenously, and this is the best biochemical evidence of deficiency. If linoleic acid is the principal dietary polyunsaturated acid, the tissue lipids will contain high proportions of ω6 acids, and if linolenic acid is the major polyunsaturated acid of the diet, the tissue lipids will be rich in ω3 acids. The pattern of polyunsaturated acids is thus useful to diagnose the EFA status of an individual. The ratio of eicosa-5,8,11-trienoic acid to eicosa-5,8,11,14-tetraenoic acid, or more simply, the ratio of total trienoic acids to total tetraenoic acids, has been useful in assessing nutritive status with respect to linoleic acid, which is the principal polyunsaturated acid in most diets (8, 9). The influence of dietary linoleate on the triene/tetraene ratio is shown in Fig. 2 for serum lipids of infants (10) and for plasma lipids of rats (8), illustrating the similarity of the biochemical phenomena of deficiency in the two species.

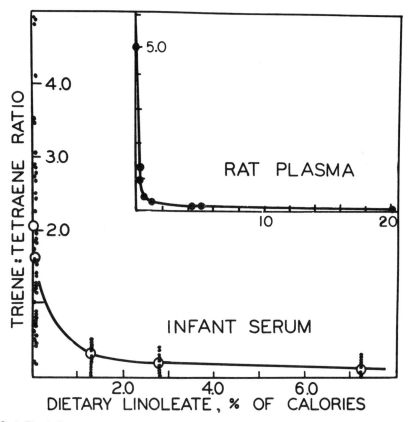

FIG. 2. The influence of level of dietary linoleic acid on the triene/tetraene ratio in tissue lipids in serum lipids of infants (10) and in plasma lipids of rats (8).

Hansen and his group (11) studied the fatty acid composition of serum lipids of 428 infants fed formula diets the lipid and fatty acid compositions of which were known. They deduced empirically that the requirement for the infant was on the order of 1% of calories of linoleic acid. Plotting their data as triene/tetraene ratio against linoleic acid intake in percent of calories confirmed their conclusion (10). Using the same data, logarithmic regression equations were set up relating polyunsaturated acids of serum lipids to the linoleate intake (10), and the following relationship was found:

$$\log_{10} \text{linoleate (cal-\%)} = -1.087 + 0.0432 \text{ (diene} - \text{triene} + \text{tetraene)}$$

This relationship has a correlation coefficient of $r = 0.89$ between actual and calculated dietary linoleate. When one considers that the nutritionist usually desires merely an estimate which will distinguish deficiency from normalcy, this relationship is precise enough, and should be useful clinically. A similar relationship has been deduced for young adult males (12).

The investigations of Hansen and his colleagues revealed that some infant formula diets that were used at that time were deficient in essential fatty acids. In fact, skin lesions were induced in a few infants given a diet com-

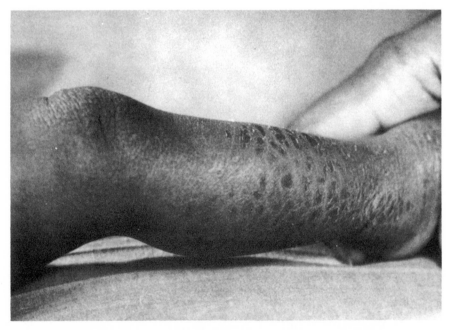

FIG. 3. Dermatitis of EFA deficiency in an infant. (Courtesy of A. E. Hansen and H. F. Wiese.)

pounded with skim milk and which thus contained very little linoleate (13). An example of the dermatitis of EFA deficiency in infants is shown in Fig. 3. Dry scaly skin was also observed in a child who had chylous ascites, and who was given a low-fat diet in an attempt to minimize the ascites (14). In this case, an extremely low level of polyunsaturated acids was found in the total serum lipids, and the triene/tetraene ratio measured serially in time ranged from 0.37 to 2.55, with an average of 1.23.

The newborn infant has a pattern of polyunsaturated acids in its serum lipids which resembles that of marginal EFA deficiency (15). The triene/tetraene ratio for fatty acids from serum lipids of nine newborn infants, prior to first feeding, was found to be 0.38 ± 0.11. This value is found on the curve (Fig. 2) at about 1% of calories as linoleic acid, about equal to the minimum requirement for linoleic acid. After the infant begins eating its natural food, this ratio remains low, but if it is given cow's milk as sole source of food, the ratio rises progressively into the range of deficiency (Fig. 4). It is apparent from these and other data (16) that cow's milk

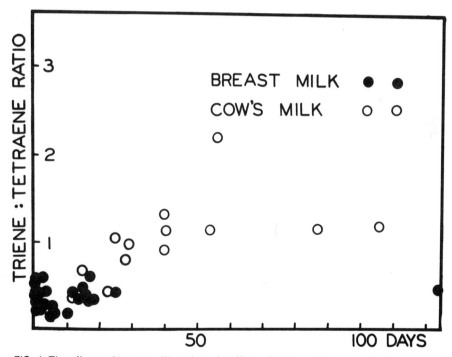

FIG. 4. The effects of breast milk and cow's milk on the triene/tetraene ratios in serum of infants (15).

formula does not provide sufficient essential fatty acids to maintain a normally low triene/tetraene ratio. It is for this reason that infant formulas now often are prepared with vegetable oil as part or all of the fat component.

Elevation of the linoleic acid content of an infant formula by incorporating an oil rich in linoleic acid does modify the composition of the fatty acids of serum lipids of infants. A modified milk containing cottonseed oil, coconut fat, and skim milk powder, having a linoleic acid content equal to 0.7% of calories, was compared with breast milk and with cow's milk formula (15). Table 1 shows that newborn infants given breast milk increased their content of dienoic and tetraenoic acids in serum lipids. Infants fed the filled milk increased the dienoic acid content even more, whereas the tetraenoic acid content remained the same as in the newborn. Breast milk or filled milk increased the trienoic acid content of serum lipids approximately twofold, whereas cow's milk formula increased the triene nearly fourfold. Hansen et al. (17) reported that the linoleic acid content of serum lipids was increased, but that of arachidonic acid was decreased, when infants were fed cow's milk formula containing 7.2% of calories of linoleic acid. Mendy et al. (18) fed a similar diet containing 5.5% of calories of linoleic acid, but found both linoleic and arachidonic acids to rise in serum lipids. Olegård and Svennerholm (19) attempted to resolve these discrepancies by a very thorough study of fatty acid composition of serum lipids of infants fed breast milk or a formula containing 5.4% of calories of linoleic acid. They found no significant difference between levels of plasma lipids in the two groups, but found that triglycerides were 25% higher in the serum of formula-fed infants. The contents of polyunsaturated acids of both plasma and red cell phospholipids were lower in the infants fed breast milk, and the difference was due solely to the $\omega 6$ acids. No increase in $20:3\omega 9$ was noted. Thus, the filled milk containing 5.4% of calories of linoleic did indeed provide more metabolically available linoleic acid than did the breast milk, the measured content of which was 3.5% of calories of linoleic acid.

TABLE 1. *The effects of breast milk, cow's milk, and filled milk on the polyunsaturated fatty acid contents in serum of infants*

	Average days	PUFA (mg/100 ml serum)		
		Diene	Triene	Tetraene
Newborn (9)	0	10.7 ± 2.3	2.7 ± 1.0	7.1 ± 1.6
Breast milk (20)	20	25.3 ± 14.9	5.3 ± 3.1	13.6 ± 5.6
Cow's milk (13)	42	18.7 ± 10.2	10.1 ± 7.1	13.3 ± 7.9
Filled milk (12)	21	44.8 ± 15.8	5.4 ± 2.3	6.2 ± 2.8

From R. T. Holman, H. W. Hayes, A. Rinne, and L. Soderhjelm (1965): *Acta Paediat. Scand.*, 54:573.

III. EFA Deficiency in Humans Induced by Intravenous Feeding

The modern practice of intravenous feeding has proven to be a life-saving measure for a large number of patients unable to take nourishment by the alimentary route. This subject has been recently discussed thoroughly in a symposium on intravenous alimentation which is being published soon (20). Most preparations used for this purpose contain protein hydrolysates, glucose, minerals, and vitamins, but no fat. Fat emulsions, although used widely previously, came into disfavor because of adverse reactions which occurred unpredictably and with sufficient frequency to be regarded as a hazard. The use of fat emulsions for intravenous feeding has been reviewed by Freeman (21), and more recently by Beisbarth and Fekl (22).

Very recently, a number of reports have indicated that the prolonged use of fat-free intravenous feeding induces an EFA deficiency. The first report of this phenomenon by Collins et al. (23) was of a 44-year-old male who had all his small bowel removed except for 60 cm, and who was maintained by fat-free intravenous feeding. They observed in this patient the advance of a skin rash coincident with development of a pattern of fatty acids in serum lipids characteristic of EFA deficiency. The rash and the abnormal pattern of fatty acids receded when intravenous fat emulsion (containing linoleate) was given, and advanced again when it was withdrawn. This is shown in Fig. 5, taken from their data. By structural analysis, they proved that the trienoic acid which increased during fat-free intravenous feeding was indeed the 5,8,11-eicosa-trienoic acid known to increase in tissue lipids in EFA deficiency. During the stage of deficiency, this acid reached 10% of the fatty acids of serum phospholipids in the patient. When a soybean fat emulsion containing 86 g of linoleic acid per liter was administered, the serum triglycerides subsided to normal, and the proportion of triglycerides of the lipoproteins increased to normal values. These investigators concluded that an adult man requires at least 7.5 g of linoleic acid per day. They also reported a single analysis of the plasma of a 61-year-old woman who had failed to thrive and had several hospital admissions for malabsorption over a period of many years because of a total colectomy. This patient's plasma phospholipid fatty acids contained 4.2% of eicosatrienoic acid, plasma triglycerides were above normal, but only 8% were present in lipoproteins of density less than 1.006.

Caldwell et al. have also observed EFA deficiency in an infant (24) in whom a midgut volvulus required resection of 50% of her small bowel at 10 days of age. Several attempts to anastomose the duodenum to the remaining small bowel failed, leaving her with a duodenostomy and ileostomy. Intravenous feeding was begun on the 18th day and continued for half a

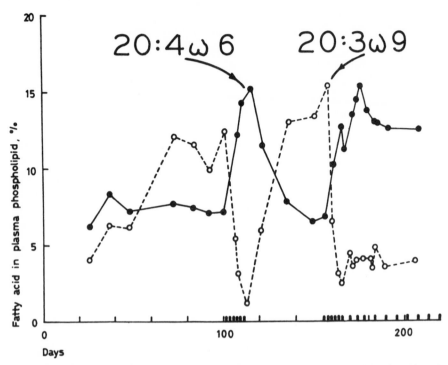

FIG. 5. The effects of fat-free intravenous feeding and of soybean oil emulsion introduced intravenously on the eicosatrienoic acid and eicosatetraenoic acid contents of plasma phospholipids in a 44-year-old man (23).

year. During this period, several attempts to close an enterocutaneous fistula failed because of poor wound healing (Fig. 6). Scaly skin, sparse hair growth, and thrombocytopenia were observed after 17 weeks. At 22 weeks, analysis of the plasma fatty acid levels revealed the pattern characteristic of the state of EFA deficiency. At 25 weeks, intravenous administration of Intralipid® was begun via periferal vein at a level to provide 4% of daily calories as linoleic acid. Linoleic acid and arachidonic acid contents of total plasma fatty acids increased markedly, and 5,8,11-eicosa-trienoic acid decreased. These changes were concurrent with healing of the skin lesions and correction of the thrombocytopenia. At 28 weeks, the duodenum and the ileum were joined and the enterocutaneous fistula was closed with complete healing of all surgical wounds. The patient was later discharged on a normal oral diet.

Caldwell et al. also considered the use of serum infusions to provide essential fatty acids to patients maintained on intravenous feeding. They

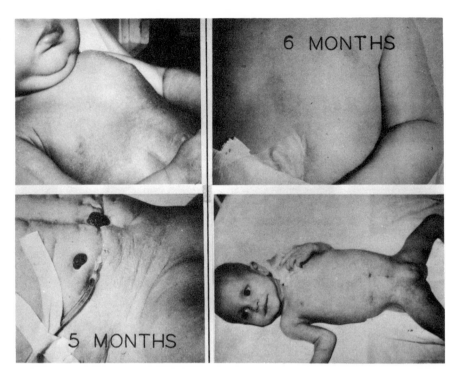

FIG. 6. Infant studied by Calwell et al. (24) showing (left) scaliness of skin and poor wound healing at 5 months, and (right) normal skin and healed wounds at 6 months, after Intralipid ® had been administered.

calculated that it is not possible to provide the necessary 4% of total calories as essential fatty acids by periodic infusions with plasma or whole blood. Their patient would have required 2800 ml of plasma daily to provide 4% of her 650 calories as linoleic acid. Thus, the use of blood or plasma is only a futile token attempt which cannot realistically meet the minimum requirement for essential fatty acids.

Belin et al. (25) have also reported on a case which developed EFA deficiency while under a course of intravenous feeding. The male infant developed a midgut volvulus at 10 days of age, and a resection was performed which left the child with a total of 7.5 cm of small intestine, a duodenostomy, and an ileostomy. He was maintained on intravenous alimentation for a year, when this was discontinued because of repeated episodes of sepsis. At 13 months, after intravenous feeding had ceased, serum lipids were still found to have the familiar pattern of EFA deficiency (Table 2).

Thereafter, the child was maintained on regular food, and iodochlorhy-droxyquin was used to control the flora of the alimentary tract. The bowel lengthened, the transit time increased, and the child is developing normally psychologically and socially. At 28 months, analysis of the serum lipids revealed that the fatty acid pattern had shifted more toward normalcy.

TABLE 2. *Fatty acid composition of serum lipids of male infant J. K. shortly after the end of 1 year of intravenous alimentation (13 months) and at end of the study (28 months) compared with average values for ten normal infants 4 months of age*

	16:1	18:1	18:2	20:3ω9	20:4ω6	20:3/20:4
Phospholipids						
13 months	2.8	20.3	9.2	4.4	8.2	0.54
28 months	1.9	13.0	21.4	1.9	8.7	0.22
10 normals	1.3	14.5	17.7	0.4	11.6	0.30
Triglycerides						
13 months	8.9	47.4	3.9	1.2	0.5	2.40
28 months	6.6	40.7	13.7	0.5	0.7	0.71
10 normals	5.0	44.8	15.4	0.2	1.0	0.2
Cholesteryl esters						
13 months	15.4	35.0	22.4	1.9	5.8	0.33
28 months	8.4	21.7	43.8	0.5	4.5	0.11
10 normals	5.0	26.3	42.5	tr	8.3	—
Free fatty acids						
13 months	—	—	—	—	—	—
28 months	—	—	—	0.3	0.9	0.33
Total lipids						
13 months	—	—	—	2.4	5.3	0.45
28 months	—	—	—	0.7	4.6	0.15

From R. P. Belin, J. D. Richardson, E. S. Medley, R. A. Beargie, L. R. Bryant, and W. O. Griffin (1972): *J. Surg. Res.*, 3:185.

In our laboratory, we have had opportunity to study serially the fatty acid changes in serum lipids of ten infants maintained on fat-free intravenous feeding. Several of these cases have been reported in preliminary fashion (26, 27), and a full report of seven cases has appeared recently (28). Infants which for various reasons required prolonged intravenous feeding were given the usual protein hydrolysate, glucose, minerals, and vitamins for various lengths of time. Serum lipids were fractionated into phospholipids, triglycerides, and cholesteryl esters, and the fatty acid compositions of each were measured by gas-liquid chromatography. When the serial analyses were plotted against time on the intravenous regimen, we found that all cases showed shifts in composition in the direction of EFA deficiency. As time increased, the deviation from normalcy increased. Although the magnitude of change was different for the different patients, the shift was always in the direction of deficiency. In Figs. 7, 8, and 9, the contents of some of

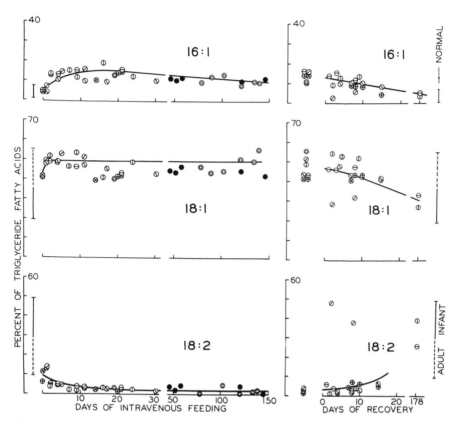

FIG. 7. The effect of prolonged intravenous feeding of a fat-free preparation and the effect of refeeding normal food on the fatty acid composition of serum triglycerides (28). Each individual is indicated by a different symbol. The black circles are values for a 77-year-old woman shown for comparison.

the fatty acids of serum triglycerides, cholesteryl esters, and phospholipids are plotted against time. The fatty acids chosen are those major components of each class which reflect the onset of EFA deficiency most dramatically.

The triglycerides (Fig. 7) contain only small amounts of fatty acids with more than 18 carbon atoms, and EFA deficiency affects only 16:1, 18:1, and 18:2 significantly. Unfortunately, samples were not available from the patients prior to the hospitalizations and intravenous therapy, so their control values are not available. However, 16:1 and 18:1 were found to increase significantly above the control values from infants of similar age. Conversely, 18:2 was found to be subnormal for all cases during intravenous feeding. For those individuals who were later given normal food, these three parameters reverted toward normal.

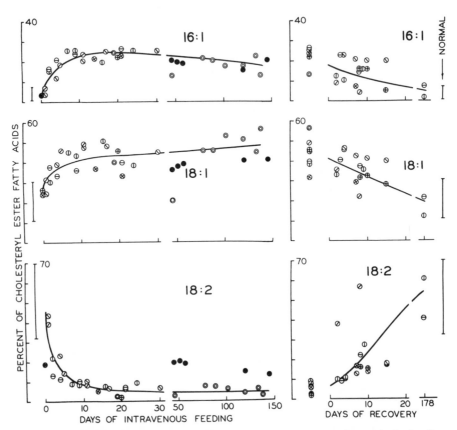

FIG. 8. The effects of intravenous feeding with a fat-free preparation and of refeeding with normal food on the fatty acid composition of serum cholesteryl esters (28).

Cholesteryl esters (Fig. 8) showed even more drastic shifts in fatty acid composition. During intravenous feeding of a fat-free preparation, 16:1 increased threefold, 18:1 increased twofold, and 18:2 dropped to one-fifth of normal values. These are the principal fatty acids of cholesteryl esters, so this represents a very thorough shift in composition. When normal feedings were restored, these values reverted toward normal.

Phospholipids (Fig. 9) revealed best the shifts in composition of the long-chain trienoic and tetraenoic acids demonstrated repeatedly to occur in EFA deficiency. The content of 18:2 (linoleic acid) diminished abruptly in a few days to a small fraction of the normal values. The decrease of its metabolically related acid, arachidonic acid (20:4ω6), was much slower, indicating that the turnover rate of 18:2ω6 is greater than the turnover rate

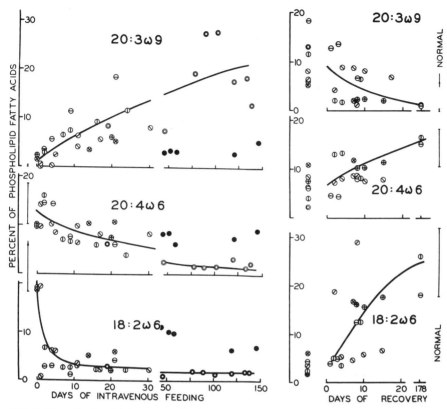

FIG. 9. The effects of intravenous feeding with a fat-free preparation and of refeeding with normal food on the fatty acid composition of serum phospholipids (28).

of 20:4ω6. Conversely, the content of 20:3ω9 increased sometimes tenfold, a strong proof of inadequate supply of essential fatty acid. These strong evidences of EFA deficiency all reverted to normal when the infants were given normal food containing essential fatty acids.

In Figs. 7, 8, and 9, the points shown in solid black represent analyses on serum samples taken from a 77-year-old woman whose small intestine was removed because of a mesenteric infarction, and who was maintained on the usual fat-free intravenous alimentation (29). The contents of the several fatty acids deviated from normal in the direction of EFA deficiency, but in many cases the degree of deviation was not as great as it was in infants fed intravenously for the same length of time. Thus, the larger stores of the older individual buffered the shift in composition due to the fat-free alimentation.

The several reports on the EFA deficiency induced by fat-free intravenous feeding indicate that a variety of physiological effects are induced in humans, as well as the biochemical changes in lipid composition. Skin lesions, thrombocytopenia, and impaired wound healing were observed in coincidence with the classical changes in fatty acid composition of serum lipids. Intravenous feeding by indwelling catheter is universally known to be associated with a high incidence of sepsis at the site of the catheter (20, 30).

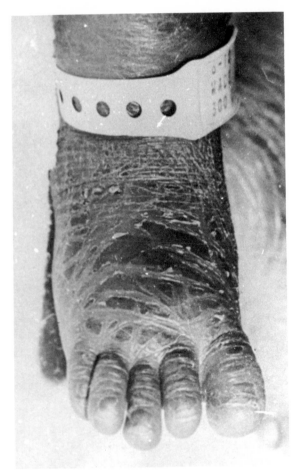

FIG. 10. Dermatitis developed in a severe case of EFA deficiency induced by fat-free intravenous feeding (28).

EFA deficiency is also known to be associated with a higher than normal incidence of infections in animals (31, 32) and in infants (11). It is not known how much, if any, the EFA deficiency contributes to the ease of invasion of the patient by microorganisms. It would seem probable that tissue of abnormal composition may be more easily invaded by organisms, and it would probably be less able to replace itself once invaded because its stores of necessary structural components for new tissue are depleted.

IV. AN EXTREME CASE OF EFA DEFICIENCY IN A HUMAN

One of the cases included in our study above, and indicated by the double circles in Figs. 7, 8, and 9, experienced a volvulus at birth and all of the small intestine except the duodenum and 2.5 cm of the terminal ileum were removed surgically. This infant was maintained for 4.5 months by intravenous alimentation, until it succumbed to meningitis. The fatty acid composition of the serum lipids indicated the most severe deviation from normal composition ever observed in this laboratory in animals or man. The triene/tetraene ratio of serum phospholipids reached its highest value of 18 at

TABLE 3. *Content of certain fatty acids in phosphatidyl choline from tissues of an EFA-deficient infant*

Tissues	16:1	18:1	18:2	20:3ω9	20:4ω6
Liver	6.9	32.6	1.1	9.9	2.0
Kidney	5.2	41.6	0.7	7.0	2.6
Pancreas	5.8	37.2	0.6	4.8	0.8
Adrenal	5.4	37.8	0.7	11.0	2.8
Colon	3.8	40.8	0.9	4.8	1.9
Ventricle	2.3	45.1	0.3	7.8	3.1
Spinal cord	5.7	41.8	0.6	1.1	1.0
Serum	5.5	30.8	1.8	13.1	2.2

From J. R. Paulsrud, L. Pensler, C. F. Whitten; S. Stewart, and R. T. Holman (1972): *Am. J. Clin. Nutr.,* 25:897.

approximately 100 days. The previously highest value observed in deficient animals was 6. During this period the infant developed a severe scaliness of skin, shown in Fig. 10. When the child died of meningitis, samples of tissues were taken at autopsy for lipid analysis. The fatty acid contents of the phosphatidyl choline isolated from each tissue are shown in Table 3. Normal samples for comparison are not available, but these compositions resemble those of lipids from EFA-deficient animals (33).

V. CONCLUSION

The amassed evidence proves that essential fatty acids are vital to human nutrition as well as to animals. In recent years the attitude that essential fatty acids are of no real meaning to human nutrition because no one could produce an overt deficiency in man has been suddenly supplanted by the realization that EFA deficiency is a common and hazardous occurrence among the gravely ill in our hospitals. It is probably no exaggeration to state that EFA deficiency is being induced unknowingly by the life-saving practice of intravenous feeding by physicians and surgeons in every major hospital in the United States and around the world. The prolonged use of the usual fat-free intravenous preparations will surely precipitate an EFA deficiency, creating new and unknown consequences and hazards, causing one nutritional deficiency while attempting to avoid others. The scientific and medical communities must become aware of this hazard and must find means of providing essential fatty acids to those patients who require long-term intravenous alimentation.

REFERENCES

1. Soderhjelm, L., Wiese, H. F., and Holman, R. T. (1971): *Progress in the Chemistry of Fats and Other Lipids,* Vol. IX. Pergamon Press, Oxford, p. 555.
2. Burr, G. O., and Burr, M. M. (1929): *J. Biol. Chem.,* 82:345.
3. Brown, W. R., Hansen, A. E., Burr, G. O., and McQuarrie, I. (1938): *J. Nutr.,* 16:511.
4. Hansen, A. E. (1933): *Proc. Soc. Exptl. Biol. Med.,* 31:160.
5. Hansen, A. E. (1937): *Am. J. Dis. Children,* 53:933.
6. Hansen, A. E., Knott, E. M., Wiese, H. F., Shaperman, E., and McQuarrie, I. (1947): *Am. J. Dis. Children,* 73:1.
7. Holman, R. T. (1971): *Progress in the Chemistry of Fats and Other Lipids,* Vol. IX. Pergamon Press, Oxford, p. 275.
8. Holman, R. T. (1960): *J. Nutr.,* 70:405.
9. Mohrhauer, H., and Holman, R. T. (1963): *J. Lipid Res.,* 4:151.
10. Holman, R. T., Caster, W. O., and Wiese, H. F. (1964): *Am. J. Clin. Nutr.,* 14:70.
11. Hansen, A. E., Wiese, H. F., Boelsche, A. N., Haggard, M. E., Adam, D. J. D., and Davis, H. (1963): *Pediatrics,* 31(Suppl. 1, pt. 2):171.
12. Holman, R. T., Caster, W. O., and Wiese, H. F., (1964): *Am. J. Clin. Nutr.,* 14:193.
13. Soderhjelm, L., and Hansen, A. E. (1962): *Pediat. Clin. N. Am.,* 9:927.
14. Warwick, W. J., Holman, R. T., Quie, P. G., and Good, R. A. (1959): *Am. J. Dis. Children,* 98:317.
15. Holman, R. T., Hayes, H. W., Rinne, A., and Soderhjelm, L. (1965): *Acta Paediat. Scand.,* 54:573.
16. Pikaar, N. A., and Fernandez, J. (1966): *Am. J. Clin. Nutr.* 19:194.
17. Hansen, A. E., Wiese, H. F., Adam, D. J. D., Boelsche, A. N., Haggard, M. E., Davis, H., Newson, W. T., and Pesut, L. (1964): *Am. J. Clin. Nutr.,* 15:11.
18. Mendy, F., Hirtz, J., Berret, R., Rio, B., and Rossier, A. (1968): *Ann. Nutr. Alim.,* 22:267.
19. Olegård, R., and Svennerholm, L. (1971): *Acta Paediat. Scand.,* 60:505.
20. Meng, H. C. *AMA Symposium on Parenteral Nutrition, in press.*

21. Freeman, S. (1955): *Progress in the Chemistry of Fats and Other Lipids,* Vol. III. Pergamon Press, Oxford, p. 1.
22. Beisbarth, H., and Fekl, W. (1971): *Z. Ernährungs Wissenchaft.,* Suppl., 10:73.
23. Collins, F. D., Sinclair, A. J., Royle, J. P., Coats, D. A., Maynard, A. T., and Leonard, R. F. (1971): *Nutr. Metab.,* 13:150.
24. Caldwell, M. D., Jonnson, H. T., and Othersen, H. B. (1972): *Pediatrics,* 81:894. See also Caldwell, M., Jonnson, J. T. and Levkoff, A. (1972): *Clin. Res.,* 20:106 (Abstract); and Jonnson, H. T., Caldwell, M. D., Levkoff, A. H., and Brueggman, J. L. (1972): *J. Am. Oil Chem. Soc.,* 49:87A (Abstract).
25. Belin, R. P., Richardson, J. D., Medley, E. S., Beargie, R. A., Bryant, L. R., and Griffin, W. O. (1972): *J. Surg. Res.,* 3:185.
26. Pensler, L., Whitten, C., Paulsrud, J., and Holman, R. T. (1970): *Abstracts of American Pediatric Society and Society for Pediatric Research,* p. 177.
27. Paulsrud, J. R., Stewart, S. E., Whitten, C. F., and Holman, R. T. (1970): *World Congress: International Society for Fat Research and the American Oil Chemists' Society,* Chicago, October 1, Abstract No. 342, p. 187.
28. Paulsrud, J. R., Pensler, L., Whitten, C. F., Stewart, S., and Holman, R. T. (1972): *Am. J. Clin. Nutr.,* 25:897.
29. Varco, R. L., Shea, M. A., Paulsrud, J. R., Stewart, S., and Holman, R. T. (1970): Unpublished data.
30. Curry, C. R., and Quie, P. G. (1971): *New Engl. J. Med.,* 285:1221.
31. Hansen, A. E., Beck, O., and Wiese, H. F. (1948) *Federation Proc.,* 7:289 (Abstract).
32. Hansen, A. E., and Wiese, H. F. (1951): *Texas Rep. Biol. Med.,* 9:491.
33. Holman, R. T. (1971): *Progress in the Chemistry of Fats and Other Lipids,* Vol. IX. Pergamon Press, Oxford, p. 275.
34. Marcel, Y. L., Christiansen, K. and Holman, R. T. (1968): *Biochim. Biophys. Acta,* 164:25. 164:25.

Dietary Lipids and Postnatal Development
Raven Press, New York © 1973

Differentiation in the Biological Activity of Polyunsaturated Fatty Acids

U. M. T. Houtsmuller

Unilever Research, Vlaardingen, The Netherlands

I. INTRODUCTION

Many animal species show signs of deficiency when raised on diets free of fatty acids. Symptoms have been observed in birds, fish, insects, and many mammal species, including man (1). The most elaborate investigations have been carried out in rats, which resulted in an extensive list of symptoms, but the full syndrome is not yet understood completely.

In addition, the question of which fatty acid(s) is (are) needed to prevent deficiency has not yet been answered. Often linoleic acid (and its biosynthetic derivatives, such as arachidonic acid) are regarded as the sole family with full essential fatty acid (EFA) potency. However, specific functions for linolenic acid have also been claimed, such as in the prevention of multiple sclerosis (2). Evidently, more elaborate studies are required before the biological function of EFA can be fully appreciated.

In this chapter, studies on the following symptoms of EFA deficiency in rats are described: decreased body weight, increased skin permeability, increased spontaneous swelling of liver mitochondria *in vitro,* and altered fatty acid pattern of liver mitochondria.

II. METHODS

The symptoms of EFA deficiency were investigated in two types of curative bioassays (Fig. 1). In the first method, according to Deuel (3), newly weaned rats are raised on a fat-free diet for about 14 weeks, causing a retardation in their growth rate. If the diet of some of the animals is then supplemented with a known amount of linoleic acid, such as by the addition of sunflower seed oil, the growth rate accelerates. A difference in body weight compared with the deficient control group ensues, which can be used as a standard for the estimation of the biological activity of an unknown fatty acid.

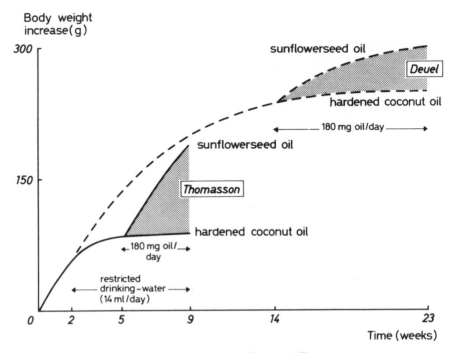

FIG. 1. Bioassays according to Deuel and Thomasson.

In the bioassay according to Thomasson (4), the rats also receive a fat-free diet, but in addition the supply of drinking water is restricted to 14 ml/day. The result is that the body weight levels off rapidly, but the same amount of sunflower seed oil causes a much larger increase in body weight than in Deuel's bioassay. This makes possible a more accurate estimation of the biological activity of other fatty acids.

III. BODY WEIGHT

Both bioassay methods were used to compare the activities of linoleic and linolenic acid (Fig. 2). The effects of a fish oil, containing very low amounts of fatty acids from the linoleic acid type (<0.8%), and of a concentrate of timnodonic (20:5, n - 3) and clupanodonic acid (22:6, n - 3), which was prepared from a fish oil, were tested at the same time.

In the Deuel assay (Fig. 2, broken lines), linoleic and linolenic acid induced about the same increase in body weight, but in the Thomasson assay, with restricted drinking water (Fig. 2, solid lines), linoleic acid was much

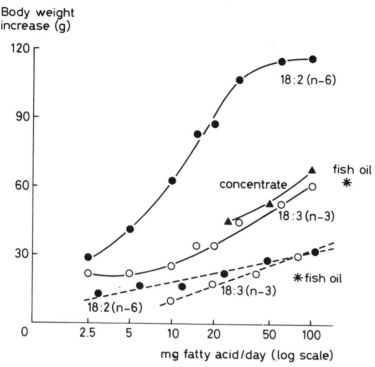

FIG. 2. Increase in body weight as a function of the fatty acid dose; bioassays according to Thomasson (solid lines) and Deuel (broken lines).

more potent than linolenic acid. With restriction of drinking water, maximal growth with linoleic acid was obtained with about 30 mg/day, whereas in the Deuel assay 100 mg/day did not seem to be optimal. Linolenic acid did not reach a maximal effect in both assays. The biological activity of linolenic acid in the Thomasson assay was about 10% with respect to linoleic acid, whereas in the Deuel assay this activity was 100%. With respect to the fish oil and the concentrate, both assays exhibited the activity which is in accordance with the content in (n - 3) fatty acids.

IV. SKIN PERMEABILITY

Obviously, the differences between the two methods are caused by the restriction in drinking water in the Thomasson assay. Therefore, it seemed relevant to look for an explanation in the increased skin permeability of the deficient animal. This criterion may be measured in three different ways (giving essentially the same results): the loss of water by the anesthetized rat, the permeation of water through an isolated piece of skin *in vitro,* and the consumption of water when freely available (not to be used in the Thomasson assay).

A linear relationship between the loss of water by the anesthetized rat

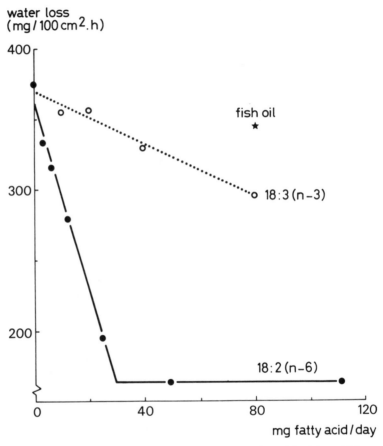

FIG. 3. Loss of water by the anesthetized rat as a function of the fatty acid dose.

and the amount of fatty acid administered was found for linoleic as well as for linolenic acid (Fig. 3). However, the slope of the line is much steeper for linoleic acid. From these differences in slope, an activity of 10% for linolenic acid can be calculated. Normal skin permeability was obtained with

30 mg/day of linoleic acid, which was also the amount that caused almost maximal growth in the Thomasson assay (Fig. 2).

For linolenic acid, on the other hand, a maximal effect was not obtained with the doses used, neither with respect to growth nor with respect to skin permeability. Therefore, it seemed appropriate to use much higher doses in a subsequent experiment. Since, however, an adequate supply of the pure fatty acids was not available, use was made in a Deuel assay of sunflower seed oil and the fish oil that had been tested in the previous experiments. The very small amount of linoleic acid in this oil apparently can be neglected. Maximal body weight was obtained for both fats with about 180 mg/day, which in the case of sunflower seed oil corresponded to 110 mg/day of linoleic acid. However, the body weight of the rats that recovered from EFA-deficiency never reached the level of the rats fed a normal diet from weaning.

Skin permeability, measured *in vitro,* is restored to normal with 50 mg/day of sunflower seed oil, which is the same amount of linoleic acid (30 mg/day) as found in the previous experiments. The skin also becomes normal with very high doses of fish oil. This effect cannot be ascribed to the small amount of (n - 6) fatty acids in this oil; the (n - 3) fatty acids must also have contributed.

The increase in body weight in the Thomasson bioassay compares very well with the improvement of the skin permeability (30 mg/day of linoleic acid is optimal in both criteria; linolenic acid has only 10% activity). In the Deuel assay, however, much higher doses (about 110 mg/day of linoleic or linolenic acid) are needed for optimal growth.

Since 14 ml/day of drinking water is low, even for normal rats, the EFA-deficient rats lose a considerable part of this water through their skin. Thus, a water deficiency ensues along with the EFA deficiency. Since tissue contains about 70% water, new tissue cannot be formed and growth stops rapidly. When linoleic acid is administered, the skin permeability decreases, so that more water can be retained in the animal and growth becomes possible. Optimal growth will be attained at that dose of linoleic acid which completely cures the skin. Linolenic acid, which has only 10% of the potency of linoleic acid in curing the skin, will allow only a limited growth. This is what was actually observed in the bioassay of Thomasson. Moreover, in the Thomasson bioassay the change in mean body weight of the groups appeared to be a function of the food consumption, which again is a function of the amount of linoleic acid administered to the rats, but only up to 30 mg/day. Apparently, renewed synthesis of tissue, made possible by the sparing effect of linoleic acid on available water, can only be effectuated if more food is consumed.

V. SWELLING OF LIVER MITOCHONDRIA

The slight additional growth (Fig. 2), at constant food consumption, when more than 30 mg/day of linoleic acid is administered can apparently not be attributed to the water-sparing effect. Neither can the growth in the Deuel assay be thus explained, since water is available *ad libitum* and food consumption is the same for all groups. The increase in body weight is rather the result of increased food efficiency, which means that the body needs less

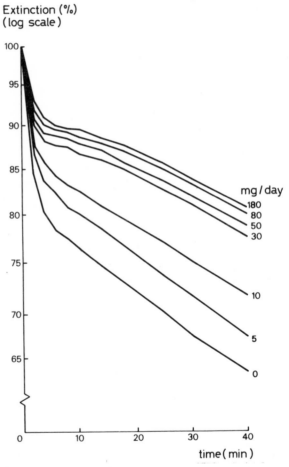

FIG. 4. Spontaneous swelling of liver mitochondria after feeding various doses of sunflower seed oil for 28 days to EFA-deficient rats.

substrate for energy production. Underlying this phenomenon may be a slight uncoupling of mitochondria. The increased swelling properties of isolated liver mitochondria might be relevant in this respect, since swelling is usually accompanied by uncoupling. As shown in Fig. 4, the spontaneous swelling of the liver mitochondria, measured as the change in optical density, is proportional to the amount of linoleic acid administered to the rats.

It has been shown (5) that the swelling curves can be described by $E_t(\%) = Ae^{-\alpha t} + Be^{-\beta t}$, in which $E_t(\%)$ = optical density at time t, expressed as a percentage of the initial value, and A, B, α, and β are parameters which are correlated with the amount of linoleic acid fed to the rats. For practical purposes, only α has been used as a criterion of EFA deficiency. When α is plotted as a function of the dose of linoleic and linolenic acid, both show the same linear correlation. At a dose of 100 mg/day, the optimal effect is not yet reached. The fish oil and the concentrate also fit well within the correlation. Therefore, this criterion parallels in every respect the increase in body weight in the Deuel assay, which suggests that the latter is due to changes in mitochondrial properties, resulting in better food efficiency. In contrast, the Thomasson bioassay mainly measures the skin permeability, but some effect of improved food efficiency is also observed.

VI. FATTY ACID PATTERN

It is well known that in EFA deficiency the fatty acids of the linoleic acid family are replaced by those of the oleic acid group, particularly arachidonic acid by eicosatrienoic acid. Holman (6) has even introduced the triene/ tetraene ratio as a criterion of EFA deficiency. This criterion implies that arachidonic acid has properties that cannot be met by eicosatrienoic acid. These properties may either be structural, such as in phospholipids, or dynamic, such as the production of prostaglandins.

In mitochondria a structural function seems preponderant, since these organelles have a high content of polyunsaturated fatty acids in their membranes. In the present experiments the replacement of eicosatrienoic acid in the liver mitochondria has been studied (Fig. 5).

The fatty acids of the (n - 3) family are incorporated three times more readily than the (n - 6) fatty acids, and the mitochondria have even higher "affinity" for the fatty acids in the concentrate. However, in normalizing the swelling properties of the mitochondria, linoleic and linolenic acid showed the same activity. This means that three times as many fatty acids of the linolenic acid type are needed for the same structural effect, or that in this criterion linolenic acid has 30% of the biological activity of linoleic acid.

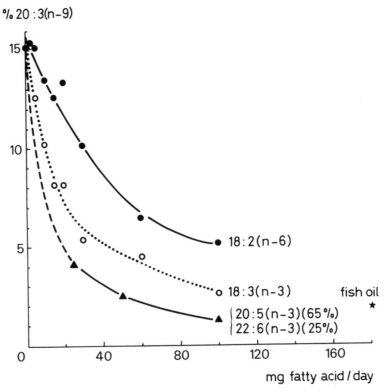

FIG. 5. Replacement of 20:3 (*n* - 9) by various fatty acids in the liver mitochondria of EFA-deficient rats.

VII. OTHER POLYUNSATURATED FATTY ACIDS

A series of synthetic polyunsaturated fatty acids (synthesized in our department of organic chemistry) of variable chain length and different degrees of unsaturation were tested for EFA activity in bioassays with restricted drinking water. At the end of the feeding period the mitochondrial fatty acid pattern and swelling properties were established, and in some instances also the skin permeability *in vitro* (Fig. 6). Only 22:3 (*n* - 8), 22:4 (*n* - 8), and 18:3 (*n* - 4) show no activity at all. The other fatty acids all normalize the swelling properties of the mitochondria to a greater or lesser extent. With respect to growth, only three fatty acids show an appreciable activity: arachidonic acid, its isomer 20:4 (*n* - 6, 9, 12, *16*), and 21:4 (*n* - 7). The remaining fatty acids have minor or no activity. In the

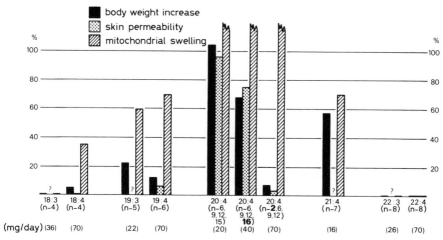

FIG. 6. EFA activities on body weight increase, skin permeability, and mitochondrial swelling of different synthetic fatty acids (percent compared with that of linoleic acid).

assays with restricted drinking water, the skin permeability again parallels growth: arachidonic acid and the (n - 6, 9, 12, 16) isomer have high potency, the third isomer has low potency. Unfortunately, the skin permeability of 21:4 (n - 7) was not estimated, but it can be inferred from the high growth rate that this fatty acid decreases the loss of water. Possibly 19:3 (n - 5) also has some activity (not measured), but the other fatty acids hardly affect the skin.

When the incorporation of these fatty acids is taken into account (Fig. 7), it becomes apparent that the lack of activity of 22:3 (n - 8) is the consequence of the fact that this fatty acid is not incorporated into the mitochondria. On the other hand, 22:4 (n - 8) is readily incorporated, but still has no effect on the mitochondria. An analogous situation is found for 18:3 (n - 4) and 18:4 (n - 4), in that both are incorporated but only the tetraene shows a minor effect of swelling. The 20:4 (n - 2, 6, 9, 12) isomer exhibits an incorporation comparable to linolenic acid, whereas the two other isomers resemble linoleic acid. As shown before, the activities of 20:4 (n - 2, 6, 9, 12) with respect to the other criteria also resembled that of the (n - 3) family. Both 19:3 (n - 5) and 19:4 (n - 6) are incorporated like linoleic acid is.

Using the amount of 20:3 (n - 9) in the tissue as a criterion of incorporation of fatty acids being tested, the "affinity" of the mitochondria for the fatty acids may be calculated (Fig. 8, left-hand side). From this and from the potency of the fatty acids to normalize mitochondrial swelling, the

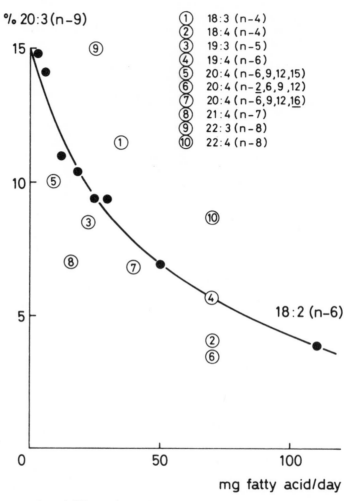

FIG. 7. Incorporation of different fatty acids into mitochondria indicated by the amount of 20:3 (n - 9) as a function of the fatty acid dose.

"effective activity" can be derived (Fig. 8, right-hand side). This "effective activity" expresses the suitability of the fatty acid to function as a structural element in the mitochondria. From 18:3 (n - 8) no value can be calculated, because this fatty acid did not incorporate. 19:4 (n - 6) is almost as well suited for the mitochondria as linoleic acid. Arachidonic acid and both its isomers, as well as the fatty acids of the fish oil concentrate, are even more adequate than linoleic acid, but the actual values were too high to be calculated accurately.

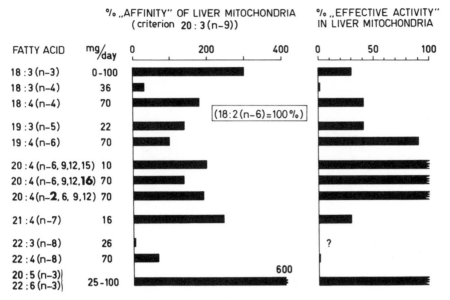

FIG. 8. "Affinity" of liver mitochondria for different fatty acids and the "effective activity" of these fatty acids in liver mitochondria.

VIII. CONCLUSIONS

The various criteria require different amounts of linoleic acid; whereas 100 mg/day was needed for optimal growth or normalization of mitochondrial swelling, only 30 mg/day was necessary for curing the skin.

Many, but not all, polyunsaturated fatty acids may replace linoleic acid as regards body weight increase and swelling of liver mitochondria.

Only a few fatty acids can restore the increased skin permeability to normal.

The incorporation of polyunsaturated fatty acids into liver mitochondria diverges widely.

Therefore, when the correlation between structure and function of a fatty acid is investigated, the effect has to be related to the extent of incorporation.

REFERENCES

1. Holman, R. T. (1968): In: *Progress in the Chemistry of Fats and Other Lipids*, edited by R. T. Holman. Pergamon Press, Elmsford, N.Y.
2. Bernsohn, L., and Stephanides, L. M. (1967): *Nature*, 215:821.
3. Deuel, H. J. (1951): *J. Nutri.*, 45:535.
4. Thomasson, H. J. (1953): *Intern. Z. Vitaminforsch.*, 25:62.
5. Houtsmuller, U. M. T., Van der Beek, A., and Zaalberg, J. (1969): *Lipids*, 4:571.
6. Holman, R. T. (1960): *J. Nutri.*, 70:405.

Dietary Lipids and Postnatal Development
Raven Press, New York © 1973

Vitamin A Transport in Man and the Rat

DeWitt S. Goodman

*Department of Medicine, Columbia University College of Physicians & Surgeons,
New York, New York 10032*

Vitamin A is a substance necessary for normal prenatal and postnatal growth and development. In the absence of vitamin A in the diet, weanling rats cease to grow, and in time die. Although the mechanism of the action of vitamin A is not understood, except for its role in vision, it is clear that many tissues require a continued supply of the vitamin to sustain normal differentiated function.

Vitamin A is obtained in the diet either as the preformed vitamin or as a provitamin A compound such as β-carotene. In either case, retinyl esters are formed in the intestine and are absorbed into the body via the lymphatic pathway in association with lymph chylomicrons (1, 2). The chylomicron vitamin A is removed from the circulation by the liver (3), where vitamin A is normally stored for later use. The vitamin is then mobilized from the liver as the free alcohol retinol, which is bound to a specific transport protein, retinol-binding protein (RBP). This is the form in which vitamin A is transported from the liver to the peripheral tissues, such as the eye, intestine, salivary glands, and gonads, to supply their metabolic needs. It is likely that the delivery of vitamin A to these tissues is controlled by processes which regulate the production and secretion of RBP by the liver.

I. HUMAN RBP AND PREALBUMIN

The transport system for vitamin A in human plasma comprises two proteins, RBP and prealbumin (4, 5). RBP has a molecular weight of approximately 21,000, α_1 mobility on electrophoresis, and a single binding site for one molecule of retinol. RBP interacts strongly with prealbumin, and normally circulates in plasma as a 1:1 molar RBP-prealbumin complex (4–6). RBP normally circulates mainly as the holoprotein, containing a molecule of bound retinol. The usual level of RBP in plasma is about 40 to 50 μg/ml, and that of prealbumin is about 200 to 300 μg/ml (7, 8).

In addition to its role in vitamin A transport, prealbumin has been recognized for some time as one of the three proteins involved in the plasma transport of thyroid hormone (9–11). The prealbumin molecule appears to contain one major binding site for one molecule of thyroxine (12, 13). Estimates of the association constant for the binding of thyroxine to prealbumin have varied from 1.6×10^7 (13) to 2.3 or 3.6×10^8 (14, 15). Recent studies from several laboratories have indicated that the prealbumin molecule is a tetramer, composed of four very similar, and possibly, identical subunits (16–20). X-ray crystallographic studies of prealbumin are in progress (17). The amino terminal amino acid sequence of RBP is unrelated to that of prealbumin (16).

II. STRUCTURAL AND PHYSICAL CHEMICAL STUDIES

Information is available about the RBP-prealbumin protein-protein interaction, and about the interaction of each protein with its ligand. The interaction of prealbumin with thyroxine appears to be independent of the RBP-prealbumin interaction (13). In contrast, the interaction of retinol with RBP appears to be stabilized by the formation of the RBP-prealbumin complex. Thus, retinol is much less readily removed from RBP, by gentle extraction with heptane, in the presence than in the absence of prealbumin (21, 22). Moreover, fluorescence studies have demonstrated that formation of the RBP-prealbumin complex substantially increases the thermal stability of the retinol-RBP complex (23).

The techniques of circular dichroism and optical rotatory dispersion have been used to examine the effects of the interactions involved in retinol transport on the secondary structures of both RBP and prealbumin (24, 25). RBP appears to have a relatively high content of unordered conformation, a significant but small complement of β structure, and little or no α helix. Prealbumin appears to have a higher content of β conformation (one-third to one-half the overall structure) and a lower content of unordered structure than does RBP; prealbumin also appears to have little or no α-helical structure. It is likely that the structures of both proteins are highly organized and specific, but lack regular repeating characteristics. There was no evidence from the CD spectra to indicate that the interaction of either RBP with retinol or of prealbumin with thyroxine resulted in significant changes in the secondary structures of either protein. Moreover, the spectra of mixtures of RBP and prealbumin were additive, indicating that formation of the RBP-prealbumin complex results in very little if any alteration in the secondary structure of the two proteins.

The interaction of retinol with RBP is of considerable physiological importance, since this interaction serves to solubilize the water-insoluble retinol molecule and also to protect the unstable retinol molecule against chemical degradation. Retinol bound to RBP is unavailable for oxidation by liver alcohol dehydrogenase, in contrast to retinol more weakly complexed with either bovine serum albumin or β-lactoglobulin (26). The interaction of RBP with prealbumin serves the physiological role of protecting the relatively small RBP molecule against glomerular filtration and subsequent renal catabolism and/or excretion. Although the proportion of RBP in plasma present in the free state, not complexed with prealbumin, is normally very small, it is sufficient to permit a significant amount of RBP to be filtered by the glomeruli and metabolized by the kidneys each day (8).

III. CLINICAL STUDIES

Studies have been conducted to examine the effects of a variety of diseases on the plasma levels of RBP and prealbumin in man. These studies employed a radio-immunoassay specific for human RBP (7), and a radial gel diffusion immunoassay for prealbumin (8). In normal subjects the plasma levels of both vitamin A and prealbumin were significantly correlated with the levels of RBP. In patients with liver disease the levels of vitamin A, RBP, and prealbumin were all markedly decreased, and were highly significantly correlated with each other over a wide range of concentrations (8). Nineteen patients with acute hepatitis were studied with serial samples; as the disease improved, the plasma concentrations of vitamin A, RBP, and prealbumin all increased. In these patients the RBP concentrations correlated negatively with the values of standard tests of liver function (plasma bilirubin, glutamic-oxaloacetic transaminase, and alkaline phosphatase). In hyperthyroid patients both RBP and prealbumin concentrations were significantly lower than normal. In hypothyroidism neither protein showed levels significantly different from normal. In both liver and thyroid disease the molar ratios of RBP to prealbumin and of RBP to vitamin A were not significantly different from normal. In contrast, in patients with chronic renal disease the levels of both RBP and vitamin A were greatly elevated, although the prealbumin levels remained normal. The molar ratios of RBP to prealbumin and of RBP to vitamin A were both markedly elevated. The kidneys appear to play an important role in the normal metabolism of RBP.

Another disease in which we are interested is cystic fibrosis of the pancreas. When oral vitamin A is given to patients with cystic fibrosis in doses adequate to maintain normal hepatic stores, plasma levels of vitamin A remain low (27). The plasma vitamin A transport system was studied in 43 patients with cystic fibrosis receiving oral supplements of vitamin A and in 95 normal control subjects of a similar age range (28). The mean plasma concentrations of vitamin A, RBP, and prealbumin were significantly lower in the patients than in the controls. Moreover, in the cystic fibrosis patients each component of the transport system failed to show the normal age-related rise. It is not known whether these abnormalities in the retinol transport system are primary or secondary features of cystic fibrosis; the abnormalities may, however, play a role in the pathophysiology of this disease.

IV. RETINOL TRANSPORT IN RAT PLASMA

Recent studies in our laboratory demonstrated that a similar transport system for vitamin A exists in rat plasma (29). RBP was isolated from rat serum by a sequence of procedures which included precipitation with ammonium sulfate between 30 and 50% saturation, chromatography on DEAE-Sephadex, gel filtration on Sephadex G-200 and G-100, and preparative polyacrylamide gel electrophoresis. RBP was obtained which had been purified approximately 2300-fold and which was pure by physical and immunological criteria. Purified rat RBP has a molecular weight of approximately 20,000, and a sedimentation constant of 2.06 S. It has α_1 mobility on electrophoresis and apparently one binding site for one molecule of retinol. The properties of rat RBP resemble those of human plasma RBP in many ways. The two proteins have nearly identical ultraviolet absorption spectra (peak maxima at 280 nm and 330 nm) and fluorescence emission and excitation spectra. The amino acid compositions of rat and human RBP are somewhat similar, both having a fairly high content of aromatic amino acids. Nevertheless, despite these similarities, there was no immunological cross-reactivity between rat and human RBP, indicating that the two proteins are immunologically completely distinct.

In plasma, rat RBP circulates in the form of a protein-protein complex with an apparent molecular weight of approximately 60,000 to 70,000. The protein (prealbumin-2) with which rat RBP interacts to form a complex has an electrophoretic mobility slightly greater than that of rat serum albumin. Purified prealbumin-2 has an apparent molecular weight of approximately 45,000. Prealbumin-2 may also represent a major transport protein for thyroid hormone in the rat.

V. REGULATION OF RBP METABOLISM BY NUTRITIONAL VITAMIN A STATUS

A study was carried out to examine the role of dietary vitamin A in regulating the metabolism of RBP in the rat (30). The study was designed to explore in detail the effects of vitamin A depletion and deficiency, and of repletion, on the level of serum RBP. The study employed a sensitive and specific radio-immunoassay which accurately measures rat RBP in amounts of 0.5 to 2 ng per assay tube (30). Weanling rats were divided into five groups, each was fed a vitamin A-deficient diet with or without supplements as follows: Group 1: control, supplemented with vitamin A; Group 2: pair-fed control, supplemented with vitamin A but with food intake matched to that of the deficient rats; Group 3: deficient, no vitamin A supplementation; Groups 4 and 5: retinoic acid, supplemented, after the initial depletion period, with a modest daily dose (14 or 28 μg) of retinoic acid. The rats were studied for 75 days. In the deficient group the serum vitamin A levels decreased gradually during the first 25 to 30 days of the study, from 45 ± 6 (sem) μg/dl on day 3 to levels of about 2 μg/dl. Serum RBP levels also declined during the induction of vitamin A deficiency, with a time course similar to that seen for vitamin A, but with a lag of about 3 days (from 50 $\pm$ 4 μg/ml on day 3 to 20 ± 2 μg/ml on day 27). Unlike vitamin A, however, RBP did not disappear from serum. After approximately 40 days on the deficient diet, the level of RBP became relatively stable at about 12 to 15 μg/ml, representing about 25 to 30% of the level of RBP seen on the third day. Almost all of the RBP circulating at this time was present as the apoprotein, not containing a molecule of bound retinol.

The results obtained with the retinoic acid-fed rats, who continued to grow normally, were identical with those of the deficient rats. In contrast, both the control and the pair-fed groups exhibited no major changes in either serum vitamin A or RBP levels throughout the entire study. Control rats maintained fairly constant serum vitamin A levels (64 ± 3 μg/dl, after an initial small rise). The serum vitamin A and RBP levels in the pair-fed control group were similar to those in the *ad libitum* control group.

When liver homogenates from control or from deficient rats were tested in the radio-immunoassay for rat RBP, considerable amounts of immunoreactivity were found; the immunoassay curves were, moreover, indistinguishable from those generated with pure rat RBP. In the control rats the level of immunoreactive RBP in the liver remained low and constant throughout the study, at about one-half to one-third the level found in the serum of these rats. In contrast, the level of RBP in the livers of deficient rats was four times ($p < 0.001$) that in the livers of control rats. In both control and deficient rats the immunoreactive RBP was almost entirely associated with the particulate fractions of the liver homogenates. After 20 days on the deficient diet the livers of the deficient rats contained about 750 μg of RBP more than did the livers of the control rats. This amount of RBP was equivalent to or somewhat greater than the amount of RBP which would have been needed to restore the plasma (and extracellular) compartment to a normal level.

When vitamin A was administered orally to deficient rats on day 53, a very rapid increase was seen in both serum vitamin A and RBP levels. Within 5 hr (the first time of sampling) the RBP level rose from a mean of about 14 μg/ml to 56 $\pm$ 3 μg/ml. With continued administration of vitamin A to these rats, the RBP level decreased somewhat at 29 hr and then rose and stabilized at about 50 to 52 μg/ml.

These findings suggested that vitamin A deficiency primarily interferes in some way with the secretion, rather than with the synthesis, of RBP by the liver, and that the deficient liver contains a pool of previously formed apo-RBP which can be released rapidly into the serum, as holo-RBP, when vitamin A becomes available.

VI. EFFECTS OF CHYLOMICRON VITAMIN A ON RBP METABOLISM

Having arrived at these conclusions, we set out to examine the effects of vitamin A repletion in much greater detail by carrying out a study employing chylomicrons containing newly absorbed vitamin A (J. E. Smith, Y. Muto, P. O. Milch, and DeW. S. Goodman, manuscript submitted for publication). Chylomicrons were injected intravenously into vitamin A-deficient rats, and the levels of vitamin A and RBP in serum were determined on samples collected serially from individual rats; liver-immunoreactive RBP concentrations were also determined. Chylomicrons were used so that the vitamin could be administered physiologically in the form in which it is normally absorbed. After the injection of chylomicrons containing vitamin A, a rapid increase in the serum levels of RBP and vitamin A occurred, with maximal levels seen at 2 to 4 hr. The magnitude of the response was directly related to the amount of vitamin A given, in the dose range 0 to 17 μg; maximal responses were obtained with doses of 16 μg or greater. Livers were obtained 2 hr after chylomicron injection in rats given graded amounts of vitamin A. The dose-response relationship of the increase in serum RBP was mirrored by a complementary dose-related decrease in the level of RBP in the liver. Release of RBP from liver into serum, which was a function of the amount of vitamin A given, apparently occurred. Rats pretreated with inhibitors of protein synthesis, either puromycin or cycloheximide, also showed a rapid and substantial rise in serum RBP and vitamin A levels after the injection of vitamin A. The results indicate that the increased level of RBP in serum after vitamin A injection mainly represents the release of previously formed RBP from an existing pool in the liver rather than newly synthesized protein. Thus, the secretion of RBP by the liver appears to be regulated efficiently by the availability of vitamin A at the liver cell for the formation of the retinol-RBP complex. We believe that the appearance of retinol in or on the liver cell generates some kind of a "signal" which effectively stimulates both the formation of a complex between the retinol and RBP and the secretion of the resulting holo-RBP molecule into the circulation. The nature of such a "signal" and the manner of its operation are obscure. Its exploration should, however, provide insight into the factors which control vitamin A delivery from the liver to peripheral tissues, and may also provide information about factors which regulate the secretion of other plasma proteins as well.

ACKNOWLEDGMENTS

I am grateful to the many talented colleagues who participated in the studies in our laboratory summarized in this chapter. This work was supported by grants AM-05968 and HL-14236 (SCR) from the National Institutes of Health. The author is a Career Scientist of the Health Research Council of the City of New York, currently on leave in Cambridge, England, as a Fellow of the Guggenheim Foundation.

REFERENCES

1. Huang, H. S., and Goodman, DeW. S. (1965): *J. Biol. Chem.*, 240:2839.
2. Goodman, DeW. S., Blomstrand, R., Werner, B., Huang, H. S., and Shiratori, T. (1966): *J. Clin. Invest.*, 45:1615.
3. Goodman, DeW. S., Huang, H. S., and Shiratori, T. (1965): *J. Lipid Res.*, 6:390.
4. Kanai, M., Raz, A., and Goodman, DeW. S. (1968): *J. Clin. Invest.*, 47:2025.
5. Peterson, P. A. (1971): *J. Biol. Chem.*, 246:34.
6. Raz, A., Shiratori, T., and Goodman, DeW. S. (1970): *J. Biol. Chem.*, 245:1903.
7. Smith, F. R., Raz, A., and Goodman, DeW. S. (1970): *J. Clin. Invest.*, 49:1754.
8. Smith, F. R., and Goodman, DeW. S. (1971): *J. Clin. Invest.*, 50:2426.
9. Ingbar, S. H. (1963): *J. Clin. Invest.*, 42:143.
10. Robbins, J., and Rall, J. E. (1960): *Physiol. Rev.*, 40:415.
11. Oppenheimer, J. H. (1968): *New Engl. J. Med.*, 278:1153.
12. Oppenheimer, J. H., Surks, M. I., Smith, J. C., and Squef, R. (1965): *J. Biol. Chem.*, 240:173.
13. Raz, A., and Goodman, DeW. S. (1969): *J. Biol. Chem.*, 244:3230.
14. Oppenheimer, J. H., and Surks, M. I. (1964): *J. Clin. Endocrinol. Metab.*, 24:785.
15. Woeber, K. A., and Ingbar, S. H. (1968): *J. Clin. Invest.*, 47:1710.
16. Morgan, F. J., Canfield, R. E., and Goodman, DeW. S. (1971): *Biochim. Biophys. Acta*, 236:798.
17. Blake, C. C. F., Swan, I. D. A., Rerat, C., Berthou, J., Laurent, A., and Rerat, B. (1971): *J. Mol. Biol.*, 61:217.
18. Branch, W. R., Jr., Robbins, J., and Edelhoch, H. (1971): *J. Biol. Chem.*, 246:6011.
19. Rask, L., Peterson, P. A., and Nilsson, S. F. (1971): *J. Biol. Chem.*, 246:6087.
20. Gonzalez, G., and Offord, R. E. (1971): *Biochem. J.*, 125:309.
21. Peterson, P. A. (1971): *J. Biol. Chem.*, 246:44.

22. Goodman, DeW. S., and Raz, A. (1972): *J. Lipid Res.*, 13:338.
23. Goodman, DeW. S., and Leslie, R. B. (1972): *Biochim. Biophys. Acta*, 260:670.
24. Rask, L., Peterson, P. A., and Bjork, I. (1972): *Biochemistry*, 11:264.
25. Gotto, A. M., Lux, S. E., and Goodman, DeW. S. (1972): *Biochim. Biophys. Acta*, 271: 429.
26. Futterman, S., and Heller, J. (1972): *J. Biol. Chem.*, 247:5168.
27. Underwood, B. A., and Denning, C. R. (1972): *Pediat. Res.*, 6:26.
28. Smith, F. R., Underwood, B. A., Denning, C. R., Varma, A., and Goodman, DeW. S. (1972): *J. Lab. Clin. Med.* (*in press*).
29. Muto, Y., and Goodman, DeW. S. (1972): *J. Biol. Chem.*, 247:2533.
30. Muto, Y., Smith, J. E., Milch, P. O., and Goodman, DeW. S. (1972): *J. Biol. Chem.*, 247:2542.

Dietary Lipids and Postnatal Development
Raven Press, New York © 1973

Influence of Hormones on Cell Growth

Jo Anne Brasel, Hector G. Jasper, and Myron Winick

Institute of Human Nutrition, Columbia University College of Physicians and Surgeons, New York, New York 10032

The role of hormones in RNA and protein synthesis has been studied extensively. These effects are, of course, basic to their mode of action in cell growth, but there is another aspect to their action to be presented in this report, that is, their more global effects on cell hyperplasia and/or hypertrophy. Such information depends, in part, on methodology for the measurement of cell number and the assessment of cell mass at various ages.

I. CONCEPTS OF CELL GROWTH

The constancy of deoxyribonucleic acid (DNA) content of the diploid nucleus in a given vertebrate species was first proposed by Boivin et al. (1) in 1948 and confirmed by Mirsky and Ris (2) in 1949. This information had fundamental implications for the study and documentation of patterns of cellular growth. Davidson and Leslie (3) ably review the rationale, the usefulness, and the pitfalls of monitoring DNA content in tissues as an index of cell multiplication or as a reference unit for comparing compositional changes in the cell during growth.

Although the measurement of total DNA and protein content and calculation of the number of diploid nuclei and protein/DNA or weight/DNA ratios may not provide actual or precisely accurate values for cell number or cell size, it is valid to assume that increases in total DNA content represent one aspect of growth, that due primarily to DNA replication and cell division. Increases in weight or total protein content out of proportion to increases in DNA represent another aspect of growth, that due to cell enlargement. In cases of ploidy or multinucleated cells, interpretation of data must be altered. In liver, protoplasmic mass has been shown to increase commensurately with the increase in DNA (4), so that protein/DNA ratios remain valid indices. In skeletal or cardiac muscle, one can think in terms of nuclear number and of the average mass or territory presided over by one nucleus within the larger muscle cell.

Enesco and Leblond (5) published one of the first studies reporting

165

changes in weight and in DNA content or nuclear number in the major organs and tissues of a mammalian species from before birth to adulthood. On the basis of their findings in the rat, they defined two populations of different cell growth characteristics: (1) expanding cell populations, where moderate cell enlargement accounts for a considerable portion of the growth in mass of the total organ or tissue; and (2) renewing cell populations, where cell enlargement is small or absent, and growth in mass is achieved predominantly by increases in cell number. They further concluded that cell growth can be divided into three periods: initially, cellular proliferation with little or no change in cell size (hyperplasia alone); second, continued, but reduced, proliferation with the associated onset of increases in cell size (hyperplasia and concomitant hypertrophy); and last, cessation of proliferation with continued increases in cell size (hypertrophy alone). Even though information on growth of specific cell types is not delineated by these techniques, the measurement of DNA and protein provides important information regarding growth at a cellular level not available from mass measurements alone.

Studies in the rat have documented rather precisely the temporal patterns

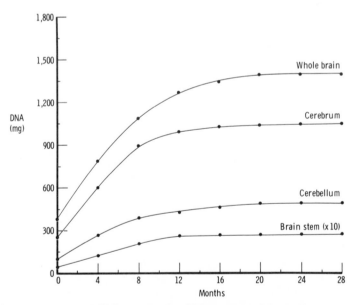

FIG. 1. Normal human brain DNA content. The DNA contents of normal human whole brain and its separate regions are plotted against postnatal age. The various regions achieve plateau, adult levels at somewhat different times and at different rates. By 12 to 18 months of age DNA synthesis is essentially complete in human brain. [Compiled from the data of M. Winick, P. Rosso, and J. Waterlow (1970): *J. Exptl. Neuro.*, 26:393.]

of the phases of cell growth for various tissues during normal growth. For example, brain and lung complete DNA synthesis by 22 days of age (6), muscle nuclear number continues to increase through adolescence (7), and DNA synthesis persists in liver, forming tetraploid and octaploid parenchymal cells, well into "middle age" (8). Much less data are available for human tissues. However, it is known that DNA synthesis ceases in human brain between 12 and 18 months of age (9) (Fig. 1) and that muscle nuclear number increases at least into adolescence in males (10). Lymphoid tissues in both species grow by an increase in cell number, with little or no increment of the average cell mass during maturation. The nonregenerating tissues, in all cases studied, continue to grow in mass and protein content, and therefore cell size, after the cessation of DNA synthesis or growth in cell number. With such data on cell growth as a background, it is then possible to study the effect of hormones on these parameters.

II. GROWTH HORMONE

In the case of pituitary dwarfism, we are particularly fortunate in having an ideal animal model. Panhypopituitarism occurs as a simple recessive

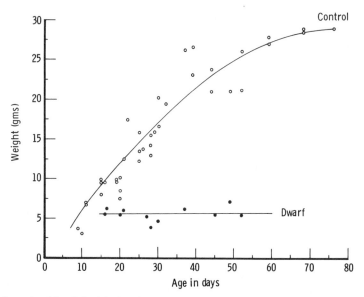

FIG. 2. Growth of Snell Smith dwarf mice. In comparison to normal littermates, the dwarf mice demonstrate considerable growth retardation first noted at 10 to 12 days of age. Prior to that time body weight measurements reveal no differences from control values. [Compiled from the data of M. Winick and P. Grant (1968): *Endocrinology*, 83:544.]

trait in the Snell Smith strain of mice. A group of these mice has been studied by Winick and Grant (11); the body weight curves are shown in Fig. 2. Before 10 to 12 days of age the dwarf cannot be identified as different from normal littermates. Soon thereafter, however, he stops gaining weight and growth essentially ceases. When one examines the tissues, the most striking changes are in DNA content, as shown in Fig. 3. Similar reductions in DNA were noted in other tissues, including muscle. These reductions in total tissue DNA content corroborate the earlier findings of Helweg-Larson (8) and others (12–15) of diminished to absent liver ploidy in these animals. Also, mitotic activity in liver and in gut mucosa have been shown to be less than normal in the absence of growth hormone (16). Utilizing radiothymi-

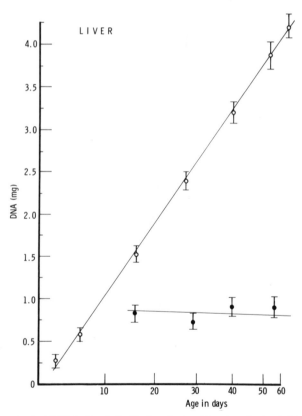

FIG. 3. Liver content of DNA in Snell Smith dwarf mice. The dwarf mice fail to demonstrate the normal increase in DNA with increasing age so that values fall further and further behind control levels with time. [Compiled from the data of M. Winick and P. Grant (1968): *Endocrinology*, 83:544.]

dine, Winick and Grant (11) have shown that DNA synthesis not only is occurring at a slower rate than normal, but also may cease prematurely in the dwarf mouse. Cheek et al. (17) have reported decreased DNA content in both liver and muscle of hypopituitary mice and rats, which increased toward normal with growth hormone therapy. In general, these same studies reveal a commensurate reduction in protein content so that protein/DNA ratios are those expected for the size of the animal. Daughaday and Reeder (18) have demonstrated that thymidine uptake in cartilage can be returned toward normal by growth hormone treatment in the hypophysectomized rat. Raben's group (19) has shown that growth hormone treatment will elevate thymidine kinase activity, an enzyme involved in the synthesis of one of the DNA precursors, in adipose tissue of hypophysectomized rats.

Very little data are available in humans. However, Cheek (20), by using creatinine to calculate muscle mass and by measuring the DNA concentra-

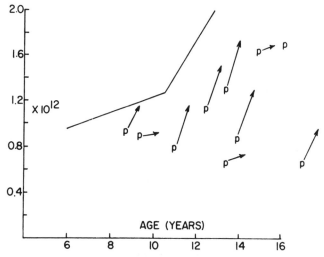

FIG. 4. Muscle nuclear number in hypopituitary dwarfs. Muscle nuclear number in males with idiopathic hypopituitary dwarfism is plotted against age. In comparison to normal male children, shown by the unbroken line, the dwarfs show considerable reduction in nuclear number prior to therapy with human growth hormone. Following 8 months of treatment, there is an accelerated rate of increase in the nuclear number, shown by the slope of the line from the initial point (p) to the tip of the arrow. However, these increases are no greater in slope than is seen for normal males during the adolescent growth spurt which begins at 10½ years of age. The three patients who did not show an increase in nuclear number could not be separated from the rest of the group on the basis of age, other tropic hormone deficiencies, or their linear growth response to therapy; their lack of response is unexplained. [Modified from D. B. Cheek (1967): Muscle cell growth in abnormal children. In: *Human Growth,* edited by D. B. Cheek. Courtesy of Lea & Febiger, Philadelphia, 1968.]

tion in a biopsy sample of gluteal muscle, was able to calculate muscle nuclear number in preadolescent male hypopituitary dwarfs before and after 8 months of growth hormone therapy. In comparison to age-matched controls, the dwarfs showed a marked reduction in total muscle DNA which was corrected toward normal at an accelerated rate by growth hormone (Fig. 4). Therefore, the limited human data support those in animals that point to DNA biosynthesis as one site of action of growth hormone on cellular growth.

III. THYROID HORMONE

The majority of studies on the effects of thyroid hormone on growth have focused on RNA and protein metabolism. However, some data are available on its effects on DNA synthesis in the young, growing animal or in tissues with continued mitotic activity throughout life. One study (21) of protein, RNA, and DNA content in young rats made hypothyroid during the second week of life revealed nearly proportionate reductions of all constituents in the tissues (Table 1). Furthermore, the tissues most sensitive to the hormone deficiency were those in the most active state of cell division at the time of the insult. For example, spleen was markedly affected, whereas brain and lung were only mildly affected. The deficient animals, therefore, at 25 days of age had disproportionately large brains and small spleens. Muscle nuclear number and total liver DNA were markedly reduced in rats 2, 5, and 7 weeks after radio-iodine administration at 1 week of age (17). In both of these studies cell size, that is, the protein/DNA ratio, was much less severely affected than cell number, that is, DNA content. Other methods of study have also suggested a role for thyroid hormone in DNA biosynthesis. Thyroid deficiency decreases mitotic activity in gut epithelium (16) and retards the rate of appearance of polyploid nuclei in liver (12, 14, 22); hormone replacement returns these abnormalities toward normal. Tri-iodothyronine-induced metamorphosis in the tadpole is associated with increased thymidine uptake into DNA in the liver.* Tri-iodothyronine administration to thyroidectomized rats enhances the mitotic index and thymidine uptake into DNA in liver (23); these changes are actinomycin D- and cycloheximide-sensitive. Finally, mean doubling time of human kidney epithelial cells in tissue culture can be reduced by 25% after exposure to exogenous thyroxin (24).

An area of intensive investigation has been the study of thyroid hormone

---

*Campbell, A. M., and Keir, H. M. (1969): The effect of tri-iodothyronine on the biosynthesis of deoxyribonucleic acid in *Rana catesbeiana* tadpoles. *Biochem. J.,* 114:38p.

and brain development. The reader is referred to the collection of papers in *Hormones in Development* (25) for a comprehensive review of work in this area. Of particular interest in the present context are the papers by Balazs et al. and Hamburgh et al. Young male rats were made hypothyroid or hyperthyroid just before or after birth, and brain growth and development were assessed. The findings, based on DNA and protein measurements and on radio-autography after thymidine injection, are that the growth of the cerebrum under these circumstances is little affected; but cerebellum, on the other hand, is significantly affected. Hormone deficiency retards the rate of DNA synthesis and prolongs the usual synthetic period, so that eventually the DNA content reaches normal adult values. Exogenous hormone administration accelerates the rate of DNA synthesis and brings it to a premature cessation, so that the final result is a cerebellum with a significant reduction in DNA content. The changes in DNA content are accompanied by persistence of the fetal external granular zone in the cerebellar cortex of the hypothyroid rats, prolongation of mitosis in these cells, and delayed migration of these cells to their final destination. The authors propose that thyroid hormone may push cells into the "differentiation" phase of growth by pulling them out of the proliferative phase. Excessive hormone may cause premature cessation of this latter phase, whereas inadequate amounts of this hormone may prolong this phase and delay differentiation. Errors in timing in either direction could then lead to permanent and/or transitory abnormalities in brain development.

In summary, thyroid hormone, at any age or stage of development, has profound effects on RNA and protein synthesis which alter the cellular content of these constituents. In addition, the hormone exerts significant effects on DNA biosynthesis during proliferative cell growth which are apt to produce permanent alterations in total DNA content and/or subsequent differentiation and development. The alterations in DNA synthesis may well be secondary to the hormone's effects on RNA and protein synthesis, since some investigations have shown them to be sensitive to actinomycin D and cycloheximide. Such studies emphasize the particular vulnerability of tissues in proliferative growth to growth-affecting stimuli and point out the necessity of employing young animals in any comprehensive study of the effects of hormones on cell growth.

IV. INSULIN

Similar to growth hormone and thyroid hormone, insulin has significant effects on RNA and protein synthesis and on amino acid transport across cell membranes. It has even been postulated that insulin exerts its primary

TABLE 1. Composition of control and hypothyroid tissues[a]

Tissue	Weight (mg)	DNA (mg)	RNA (mg)	RNA/DNA	Protein (mg)	Protein/DNA
Brain						
Control	1450 (1340–1560)	1.51 (1.31–1.91)	4.54 (3.97–5.45)	3.02 (2.08–4.00)	170 (156–182)	115 (91–139)
Hypothyroid	1240 (1170–1340)	1.42 (1.11–1.91)	3.47 (3.15–3.93)	2.57 (1.94–3.13)	151 (131–187)	109 (85–137)
Percent control	85	94	76	85	88	95
Lung						
Control	445 (420–484)	3.03 (2.28–3.37)	1.87 (1.48–2.13)	0.63 (0.44–0.86)	77 (56–90)	26 (21–31)
Hypothyroid	275 (245–305)	2.33 (1.80–2.70)	1.00 (0.89–1.14)	0.43 (0.34–0.50)	54 (36–75)	23 (16–32)
Percent control	62	77	53	68	70	90
Kidney						
Control	712 (619–845)	3.45 (3.11–3.75)	4.36 (3.38–5.10)	1.27 (0.91–1.57)	131 (102–152)	38 (27–47)
Hypothyroid	353 (301–403)	2.42 (1.76–3.08)	1.70 (1.52–1.97)	0.72 (0.55–0.86)	69 (47–87)	29 (21–40)
Percent control	50	70	39	57	53	75
Heart						
Control	275 (249–290)	0.64 (0.54–0.76)	1.04 (0.90–1.21)	1.61 (1.10–2.08)	49 (45–52)	78 (59–94)
Hypothyroid	120 (91–140)	0.43 (0.31–0.47)	0.48 (0.41–0.55)	1.20 (0.91–1.70)	21 (16–28)	48 (39–55)
Percent control	44	68	46	75	42	61
Liver						
Control	2940 (2570–3530)	6.10 (4.52–8.34)	28.00 (20.51–35.50)	4.68 (3.16–5.51)	513 (396–636)	83 (73–91)
Hypothyroid	1150 (790–1380)	2.94 (1.93–3.44)	9.65 (7.31–11.90)	3.38 (2.83–3.79)	216 (148–239)	74 (69–84)
Percent control	39	48	34	72	42	89

Muscle					
Control	318	1.17	3.69	50	150
	(268–348)	(0.95–1.52)	(2.76–4.93)	(35–64)	(128–167)
Hypothyroid	124	0.36	1.83	14	71
	(100–165)	(0.28–0.43)	(1.56–2.51)	(9–20)	(59–91)
Percent control	39	31	50	28	48
Spleen					
Control	252	2.10	0.62	77	23
	(209–307)	(1.88–2.75)	(0.53–0.69)	(61–95)	(20–26)
Hypothyroid	68	0.56	0.57	17	18
	(48–92)	(0.41–0.86)	(0.38–0.70)	(13–21)	(13–31)
Percent control	27	27	92	22	78

[a] Figures in parentheses represent the range of values. There were seven animals in the hypothyroid group and six in the control group.

[From Brasel, J. A., and Winick, M. (1970): *Growth*, 34:197.]

effects on cell size (26, 27). However, the nature of its action may well be more complicated, as suggested by other studies. Pitkin et al. (28) have shown that the fetal overgrowth accompanying maternal diabetes is associated with an increase in fetal DNA content in the rat. Placentas of human diabetics contain increased amounts of DNA (29). Microscopic analysis also suggests a true hyperplastic response (30). Moreover, the studies demonstrating little insulin effect on DNA content (26, 27) were performed either in the absence of growth hormone or in older animals. In either instance the experimental design was not such as to provide favorable circumstances for demonstrating any alterations in DNA synthesis. Additionally, Younger et al. (31) have shown a 70% increase in total liver DNA and elevations of DNA polymerase activity and thymidine incorporation within 72 hr of insulin administration to diabetic rats. Growth hormone is not an absolute requirement for this insulin-mediated response, but in its absence the response is curtailed. Therefore the majority of the data would suggest that under the appropriate conditions, insulin will affect DNA synthesis as well as RNA and protein synthesis.

V. DNA POLYMERASE STUDIES

In the above discussions of the effects of the major somatic growth-promoting hormones, several recurring features are to be noted. During the proliferative growth phase, DNA synthesis or growth in cell number will be affected with or without effects on cell size. Once this growth phase is past, the major effects are likely to be on cell size. Thus the developmental stage of the animal or tissue would appear to be a critical factor in determining the cell growth effects. The mechanisms by which these changes occur are not entirely understood. However, one pertinent area of investigation is that of DNA biosynthesis. Studies have previously been cited in which thyroid hormone and insulin produced accelerated rates of DNA synthesis accompanied by elevations of DNA polymerase activity. Recent investigations (32) have focused on the influence of growth hormone on DNA polymerase activity. This enzyme was chosen because it is felt to be involved in the final steps of DNA synthesis and may be an *ad hoc* index of proliferative cell growth. Additionally, a direct correlation between the rate of DNA synthesis and enzyme activity during normal brain growth (33) has been reported. The growth failure and diminished cell division following hypophysectomy at 21 days of age in male rats is accompanied by significant reduction in the levels of DNA polymerase activity in the liver regardless of the type of DNA primer employed (Fig. 5). Within 24 hr of growth hormone administration, enzyme activity rises, reaching a peak

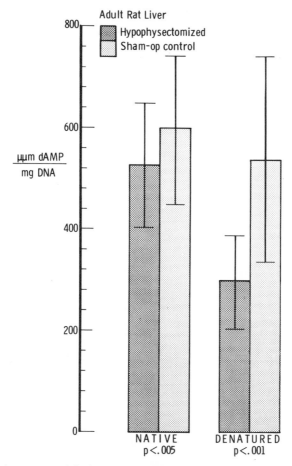

FIG. 5. DNA polymerase activity in normal and hypophysectomized rat liver. Polymerase activity is expressed in micromicromoles of dAMP incorporated per hour at 37°C per milligram of DNA for two groups of animals at 52 days of age. Male rats had previously been hypophysectomized or sham operated at 21 days of age. In the absence of the pituitary, levels of polymerase activity are reduced regardless of the primer used, but the reduction is more pronounced in that form of the enzyme employing denatured DNA as primer. Although there is overlap in the standard deviations between the groups, the large sample size of approximately 50 per group makes the differences highly statistically significant.

at 72 hr (Table 2); continued hormone administration for 2 to 3 weeks is associated with persistent elevation of the enzyme activity and liver growth which is almost entirely hyperplastic in nature, as indicated by a lack of significant change in the protein/DNA ratio (Table 3). The same dose of growth hormone produces no significant growth acceleration in normal

TABLE 2. *Effects of short-term bovine growth hormone treatment on DNA polymerase activity in hypophysectomized rats[a]*

	Treatment time (hr)				
	0	24	48	72	96
Denatured DNA primer					
Mean	295.1	569.3	647.3	945.3	554.3
S.D.	91.1	23.5	92.7	200.3	207.1
n	53	6	12	10	15
p		<0.001	<0.001	<0.001	<0.001
Native DNA primer					
Mean	524.3	836.1	722.8	1083.2	922.0
S.D.	118.6	122.4	80.7	157.3	139.0
n	64	7	8	14	13
p		<0.001	<0.001	<0.001	<0.001

[a] Activity expressed as micromicromoles of dAMP incorporated per hour at 37°C per milligram of DNA.

TABLE 3. *Effects of long-term bovine growth hormone (BGH) treatment on hypophysectomized rat liver growth*

	Without BGH	With BGH
Body weight (g)	84.0	130.8
S.D.	2.2	3.4
n	3	4
p		<0.001
Liver weight (g)	2.485	3.703
S.D.	0.083	0.328
n	3	4
p		<0.005
Liver protein (mg)	369	469
S.D.	16	20
n	3	3
p		<0.005
Liver DNA (mg)	8.1	10.0
S.D.	0.3	0.4
n	3	3
p		<0.01
Protein/DNA	45.6	46.9
S.D.	0.7	2.4
n	3	3
p		n.s.

control animals. Similar findings are noted in the kidney (Fig. 6); that is, the reduced enzyme activity noted after hypophysectomy is increased by growth hormone therapy (Fig. 7). The growth response in the first 3 weeks of therapy is nearly totally hyperplastic. By 5 to 6 weeks of treatment the response has become primarily a hypertrophic one. The transition in cell growth response occurs primarily as a result of cessation of DNA synthesis with continued growth in protein content. Concomitantly, there is a reduc-

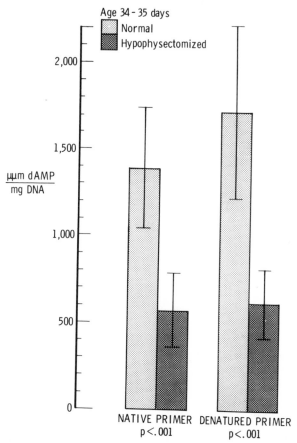

FIG. 6. DNA polymerase activity in normal and hypophysectomized rat kidney. Polymerase activity is expressed as in Fig. 5. Animals were sacrificed at 34 to 35 days of age following hypophysectomy or sham operation at 21 days. The hypophysectomized rat levels of polymerase activity are markedly reduced in comparison to control values regardless of the type of primer employed.

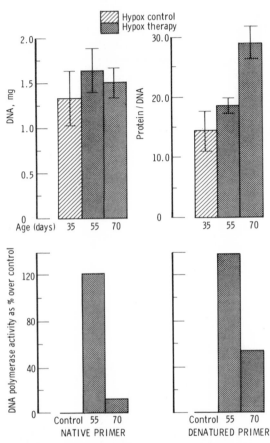

FIG. 7. Effect of bovine growth hormone treatment on the hypophysectomized rat kidney. Institution of bovine growth hormone therapy at 35 days, following hypophysectomy at 21 days, results in a significant increase in kidney DNA content at 55 days; continued therapy causes no further increase in this parameter at 70 days. Protein/DNA ratio is little changed in the first 3 weeks of therapy; however, after DNA synthesis has been completed, considerable increase in this ratio is seen by 70 days, indicating the occurrence of cell hypertrophy. DNA polymerase is expressed as the percent over age-matched, untreated, hypophysectomized control rat levels. At 55 days, in association with the increased DNA content, the treated animals show levels of polymerase activity greater than 100% above control values regardless of the primer employed. By 70 days, when DNA synthesis is no longer a prominent factor in the growth response, the enzyme levels are declining toward control values.

tion in polymerase activity as it returns towards the control levels. These data suggest that levels of DNA polymerase activity, at least, reflect the state of proliferative cell growth and that, at most, the enzyme may even represent one site of hormone action on cell growth. Whether or not alterations in polymerase activity are secondary to changes in *de novo* RNA and protein synthesis in all instances is as yet unknown.

It is also important to point out that it is nearly impossible to isolate experimentally the action of one of these hormones from the other and from changes secondary to differences in dietary intake. For example, hypophysectomy decreases both thyroid hormone and insulin secretion (34) as well as growth hormone. Hypothyroidism alters the pituitary content of growth hormone (35–38) and probably its secretion as well.* Fluctuations in blood glucose consequent to insulin deficiency or excess probably affect growth hormone secretion (39). Finally, dietary intake is almost always altered by these hormone deficiencies or excess, and altered nutrition *per se* will affect cell growth (40). This is not to suggest that it is impossible to separate the actions of specific hormones and nutritional influences from one another, since they produce distinctly different clinical and biochemical pictures: replacement of one hormone will not correct the deficits of another and pairfeeding does not produce the same picture as the hormone deficiency. However, it is certain that considerable additional information is needed to provide an overall understanding of these complex interrelationships.

Although there are many questions yet to be answered, these studies furnish a departure point for understanding the effects of hormones on cell growth. They provide a descriptive picture of the cellular make-up of tissues under states of altered hormonal status. They suggest that proliferative cell growth is particularly vulnerable to hormonal influences. Whether the effects on enzymes involved in DNA biosynthesis are specific or whether the effects are concomitant to generalized alterations in RNA and protein synthesis remains to be determined. The answers will be of interest regardless, and we hope they will provide clues to the greater question of the nature of the biochemical control points for cell growth and differentiation.

ACKNOWLEDGMENTS

This work was supported in part by the U.S. Public Health Service, National Institutes of Health. Hector G. Jasper is a recipient of a postdoctoral fellowship from the Pan American Health Organization.

---

*Sheikholislam, B. M., Lebovitz, H. E., and Stempfel, R. S., Jr. (June 21, 1966): Growth hormone secretion in hypothyroidism. *48th Annual Meeting of the Endocrine Society, Chicago.*

REFERENCES

1. Boivin, A., Vendrely, R., and Vendrely, C. (1948): *Compt. Rend. Acad. Sci.*, 226:1061.
2. Mirsky, A. E., and Ris, H. (1949): *Nature*, 163:666.
3. Davidson, J. N., and Leslie, I. (1950): *Cancer Res.*, 10:587.
4. Epstein, C. J. (1967): *Proc. Natl. Acad. Sci.*, 57:327.
5. Enesco, M., and Leblond, C. P. (1962): *J. Embryol. Exptl. Morphol.*, 10:530.
6. Winick, M., and Noble, A. (1965): *Develop. Biol.*, 12:451.
7. Cheek, D. B., Brasel, J. A., and Graystone, J. E. (1968): In: *Human Growth*, edited by D. B. Cheek. Lea & Febiger, Philadelphia.
8. Helweg-Larson, H. F. (1952): *Acta Path. Microbiol. Scand. (Kobenhavn)*, Suppl. 92.
9. Winick, M., Rosso, P., and Waterlow, J. (1970): *Exptl. Neuro.*, 26:393.
10. Cheek, D. B. (1968): In: *Human Growth*, edited by D. B. Cheek. Lea & Febiger, Philadelphia.
11. Winick, M., and Grant, P. (1968): *Endocrinology*, 83:544.
12. Carriere, R. (1955): *Anat. Rec.*, 121:273.
13. DiStefano, H. S., and Diermeier, H. F. (1956): *Proc. Soc. Exp. Biol. Med.*, 92:590.
14. Geschwind, I. I., Alfert, M., and Schooley, C. (1960): *Biol. Bull.*, 118:66.
15. Leuchtenberger, C., Helweg-Larson, H. F., and Murmanis, L. (1954): *Lab. Invest.*, 3: 245.
16. Leblond, C. P., and Carriere, R. (1955): *Endocrinology*, 56:261.
17. Cheek, D. B., Powell, G. K., and Scott, R. E. (1965): *Bull. Johns Hopkins Hosp.*, 117:306.
18. Daughaday, W. H., and Reeder, C. (1966): *J. Lab. Clin. Med.*, 68:357.
19. Epstein, S., Esanu, C., and Raben, M. S. (1969): *Biochim. Biophys. Acta*, 186:280.
20. Cheek, D. B. (1968): In: *Human Growth*, edited by D. B. Cheek. Lea & Febiger, Philadelphia.
21. Brasel, J. A., and Winick, M. (1970): *Growth*, 34:197.
22. Swartz, F. J., and Ford, J. D., Jr. (1960): *Proc. Soc. Exp. Biol. Med.*, 104:756.
23. Lee, K.-L., Sun, S.-C., and Miller, O. N. (1968): *Arch. Biochem. Biophys.*, 125:751.
24. Siegel, E., and Tobias, C. A. (1966): *Nature*, 212:1318.
25. Hamburgh, M., and Barrington, E. J. W. (1971): *Hormones in Development*. Appleton-Century-Crofts, New York.
26. Cheek, D. B., and Graystone, J. E. (1969): *Pediat. Res.*, 3:77.
27. Graystone, J. E., and Cheek, D. B. (1969): *Pediat. Res.*, 3:66.
28. Pitkin, R. M., Plank, C. J., and Filer, L. J., Jr. (1971): *Proc. Soc. Exp. Biol. Med.*, 138:163.
29. Winick, M., and Noble, A. (1967): *J. Pediat.*, 71:216.
30. Naeye, R. L. (1965): *Pediatrics*, 35:980.
31. Younger, L. R., King, J., and Steiner, D. F. (1966): *Cancer Res.*, 26:1408.
32. Jasper, H. G., and Brasel, J. A. (1973): *Endocrinology*, 92:194.
33. Brasel, J. A., and Winick, M. (1970): *Growth*, 34:197.
34. Merimee, T. J., Rabinowitz, D., Rimoin, D. L., and McKusick, V. A. (1968): *Metabolism*, 17:1005.
35. Contopoulos, A. N., Simpson, M. E., and Koneff, A. A. (1958): *Endocrinology*, 63:642.
36. Knigge, K. M. (1958): *Anat. Rec.*, 130:543.
37. Meyer, Y. N., and Evans, E. S. (1964): *Endocrinology*, 74:784.
38. Solomon, J., and Greep, R. O. (1959): *Endocrinology*, 65:158.
39. Glick, S. M., Roth, J., Yalow, R. S., and Berson, S. A. (1965): *Rec. Prog. Hormone Res.*, 21:241.
40. Winick, M., and Noble, A. (1966): *J. Nutr.*, 89:300.

Dietary Lipids and Postnatal Development
Raven Press, New York © 1973

Nutritional Effects on Brain DNA and Proteins

Myron Winick and Pedro Rosso

Institute of Human Nutrition, Columbia University, College of Physicians & Surgeons, New York, New York 10032

I. NORMAL CELLULAR GROWTH

Growth may be defined simply as an increase in weight or size of any organism or its component organs. Alternatively, growth of any organ may be defined at the cellular level as an increase in cell number, an increase in cell size, or a simultaneous increase in both. Since the DNA content is constant in all diploid cells of any species, total organ DNA content has been used to reflect cell number (1). Indeed, the actual number of cells in any organ may be calculated by determining the total DNA content of the organ and dividing by the DNA per diploid cell, a constant for the species being studied. Once the number of cells is known, a figure for the weight per cell or protein content per cell can be derived simply by dividing the organ weight or total protein content by the number of cells. This can be expressed chemically simply as a weight/DNA ratio or a protein/DNA ratio. To restate these principles: total organ DNA reflects the number of cells in the organ at the time the determination was made. The weight/DNA ratio or protein/DNA ratio reflects the average weight or protein content per cell — a measure of cell size. Similarly, the RNA/DNA ratio or the lipid/DNA ratio reflects the RNA or lipid content per cell.

By performing these chemical determinations, we are thus able to monitor growth by observing the relative contributions of increase in cell number (hyperplasia) and increase in cell size (hypertrophy).

Studies in the rat have established that DNA content reaches a maximum in all organs before growth (as measured by increase in organ weight) ceases (2). Thus, cell division stops but the organ continues to enlarge. There is a period, then, in which growth is entirely by hypertrophy. In fact, careful examination of various organs of the developing rat has revealed three phases of growth. During early growth, organ DNA content increases with total protein content increasing at the same rate; hence the ratio is un-

181

changed. Cell number is increasing, cell size remains unchanged, hyperplasia is occurring alone. Then, as a consequence of the slowing down of DNA synthesis or cell division with net protein synthesis continuing at the same rate, cell number continues to increase (although more slowly) but the ratios also begin to increase. Hyperplasia and hypertrophy are occurring together. Finally, DNA synthesis stops, indicating a cessation of cell division. Net protein synthesis still continues, the ratio increases, and hypertrophy is occurring alone.

This pattern of growth is common to all nonregenerating organs. However, the time at which cell division stops varies from organ to organ. In rat brain and lung this occurs at about 21 days of age, in heart at about 65 days of age.

II. GROWTH RETARDATION

These phases of growth have not only allowed us to define better the increase in organ size that is occurring at any given time, but they have also enabled us to predict with some degree of certainty the outcome of certain growth-retarding stimuli. For example, it has been shown that malnutrition during hyperplastic growth will retard the rate of cell division and result in an organ that is reduced in size and contains a reduced number of cells. Moreover, this change is not reversible after the period when cell division normally stops. By contrast, malnutrition during the period of hypertrophic growth prevents the increase in cell size which normally occurs and again results in a smaller organ. This change, however, is reversible at any time. The cells simply fill back up with protein, and the individual cells and entire organ achieve their normal size (3). Thus, the ability of an organ to "catch up" depends on whether cell division has been retarded. This in turn depends on when the growth-retarding stimulus has been active. The old observation that the earlier the growth failure the less likely recovery can now be explained in cellular terms.

Cell division can be curtailed in any region of an organ undergoing proliferative growth. The more rapid the rate of cell division in the area, the more profound the effects of the growth-retarding stimulus. Postnatally, for example, cell division is more rapid in rat cerebellum than in either cerebrum or brainstem (4). Malnutrition beginning at birth reduces the number of cerebellar cells more than the number of cells in other regions (5).

Any cell type undergoing division is vulnerable to the effects of malnutrition. Thus, in brain both neurons and glia will be affected (6). However, since neuronal division stops before glial division in most areas, reduction in neuronal number is a consequence of very early malnutrition.

III. MECHANISMS CONTROLLING CELLULAR GROWTH

These descriptive observations have raised some fundamental questions concerning the mechanisms controlling cell division and the ways in which certain stimuli, for example, malnutrition, affect these mechanisms. Work in this area is just beginning, but already certain important observations have been made.

Normally, as a cell grows it must continuously synthesize new protein. This synthesis takes place primarily within the cytoplasm on a subcellular organelle known as a polysome. The polysome is simply a conglomeration of varying numbers of individual ribosomes, each of which in turn is made up of a particular species of RNA. RNA itself is constantly "turning over," its quantity within the cell being determined by a balance in the rate of synthesis and the rate of degradation. The polysome "directs" the synthesis of various types of proteins by supplying templates for the various amino acids. The proteins which are synthesized may be either "structural," that is, part of the actual matrix of the cell, or "functional," that is, necessary to perform a particular cell function. The amount of structural protein synthesized, when balanced by the amount being broken down, determines the size of the cell. The amount of functional protein synthesized, when balanced by the rate at which it is degraded, determines, in part, the ability of the cell to perform certain nonspecific and specific functions. Many of these functional proteins are enzymes—proteins which catalyze specific chemical reactions. For example, certain enzymes are necessary for the synthesis and degradation of nucleic acids (both DNA and RNA) as well as for the synthesis and degradation of protein itself.

Thus, during normal growth, homeostasis is maintained in the cell as long as the raw materials are supplied in the form of amino acids and nucleotides and as long as the "machinery" for synthesizing proteins and nucleic acids functions synchronously.

Malnutrition upsets this delicate balance at several points. Evidence from a number of sources demonstrates a curtailment of DNA synthesis and protein synthesis and a reduction in RNA content. Let us examine how these changes may occur. DNA synthesis could be impaired because of lack of substrate (nucleotides), because of reduced activity of enzymes necessary to accomplish synthesis, or because of a lack of energy (ATP) for synthetic polymerization. Recent observations (7) would tend to rule out ATP lack as a crucial factor, at least in brain. We have therefore investigated the other two possibilities.

Free nucleotide per cell increases from birth to 21 days of life. Undernutrition during this period of life does not appear to affect the size of the total pool. However, since individual nucleotides have not been separated, it is possible that specific nucleotide pools are reduced. The second possible mechanism by which DNA synthesis might be affected could be by a reduction in activity of some of the enzymes involved in DNA synthesis. This could either be selective or secondary to the general reduction in protein synthesis which is known to occur.

In a previous report (8) demonstrating that activity of DNA polymerase parallels the rate of DNA synthesis in rat brain, we theorized that activity of this enzyme might serve as an *ad hoc* index of proliferative cell growth in normal tissues. Since individual brain regions have different rates of cell division and different times when maximum rates are attained, regional patterns of DNA polymerase activity also have been examined. Activity in forebrain peaks between 10 and 12 days, precisely when the rate of DNA synthesis is maximal in this region. By contrast, in cerebellum there are two peaks of DNA synthesis, at 7 and 13 days. DNA polymerase activity is also biphasic, with peaks just preceding each of the synthesis peaks. These data reinforce the concept that activity of this enzyme parallels the rate of cell division during proliferative cell growth. As such, it should provide a unique background with which to study the effects of malnutrition on DNA synthesis.

Preliminary experiments with malnutrition demonstrate a reduction in activity of this enzyme per cell in rat liver. Thus, another mechanism by which malnutrition may impair cell division could be by limiting the amount of active DNA polymerase available for DNA synthesis. Moreover, this is only one protein involved in cell division, and it is quite possible that other proteins, such as enzymes, histones, and spindle proteins, may be selectively reduced in undernutrition. The reason for this interest in protein synthesis stems from the data of others (9, 10) who have demonstrated profound quantitative and qualitative effects of undernutrition on general protein synthesis. For example, one of the earliest effects of amino acid restriction is a disaggregation of polysomes and a shift in the polysome profile, which suggests a selective sparing of certain proteins. Thus, protein synthesis which occurs on the polysome cannot really be separated from RNA metabolism. We do not know as yet where the primary effect takes place. The disaggregated polysome does not synthesize protein as well as the intact polysome. In addition, this free ribosome is much more susceptible to the activity of alkaline RNAase and hence will be broken down more rapidly.

Thus, DNA synthesis is reduced and protein synthesis is reduced either directly or secondary to changes in RNA metabolism. However, if one examines RNA metabolism, the rate of RNA synthesis actually increases during malnutrition. How can we explain this seeming paradox in view of the fact that the total amount of RNA per cell has been clearly demonstrated to be reduced? The answer must lie in enhanced catabolism. For this reason, a number of investigators have examined both the synthetic and catabolic rates of RNA metabolism during malnutrition. Studies in liver indicate an increased RNA synthesis accompanied by a proportionally higher increase in RNA breakdown, resulting in a reduced RNA per cell. Thus, the level of cellular RNA during malnutrition would appear to be controlled by a balance between synthesis and breakdown, with the regulation of this balance on the catabolic side. We have therefore decided to study RNA turnover in brain and other organs during malnutrition. To date we can report on only one aspect of the problem: the activity of alkaline RNAase.

This enzyme has been described in most tissues; it exists in a free and latent form in which the enzyme is bound to a protein inhibitor which can be released by substances interfering with SH groups such as P.C.M.B. We have measured total activity in brain during normal development and correlated this activity with changes in DNA, RNA, and protein.

Figure 1 demonstrates that RNAase activity per cell (μg DNA) increases with age except for two distinct drops at 0 and 21 days. This increase per

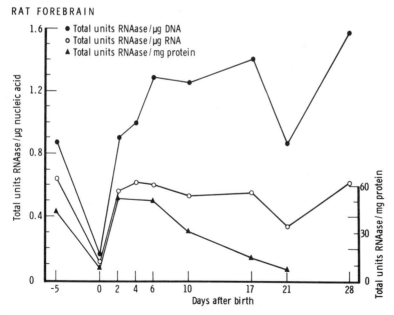

FIG. 1. Activity of alkaline RNAase during normal development in the rat forebrain.

cell is apparently not as great as the increase in other cellular proteins, and hence the activity per milligram of protein (specific activity) decreases. However, the increase exactly parallels the increase in cellular RNA content, resulting in no change in RNAase per milligram of RNA during development. Thus, a constant relationship is kept to the substrate at the same time that an increase per cell is occurring, strongly suggesting a role for this enzyme in regulating cellular RNA content. Figure 2 demonstrates that early undernutrition markedly elevates the activity of this enzyme per cell. Simi-

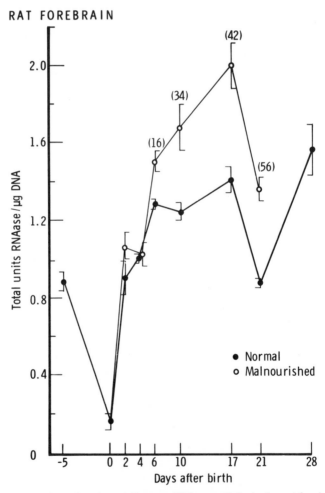

FIG. 2. The effect of undernutrition on RNAase activity in the rat forebrain.

larly, activity per milligram of RNA and protein is also elevated. Preliminary data in liver suggest similar changes with malnutrition. Thus malnutrition, which is known to enhance RNA catabolism at least in liver, elevates activity of alkaline RNAase in both liver and brain. Recently, we have studied the synthetic phase of RNA metabolism by intracerebral pulse labeling with C^{14} orotic acid. Animals malnourished from birth to 14 days of age show an increased incorporation of label both into the total nucleotide pool and into RNA during the first 24 hr after injection.

Can we fit all of the known data into a hypothesis which can interrelate the various aspects of normal cell growth and attempt to explain some of the mechanisms by which undernutrition affects this interrelation? Undernutrition redirects nucleotides away from DNA synthesis and toward RNA synthesis, thus retarding the former and increasing the latter. This retarded rate of DNA synthesis is reflected in a reduced activity of certain enzymes involved in the synthesis of DNA, for example, DNA polymerase (8). Conversely, the increased RNA synthesis appears to be accompanied by an increased activity of certain enzymes involved in RNA synthesis, for example, RNA polymerase (11).

As we have noted previously, however, total cellular RNA content decreases with malnutrition. This is a consequence of an increased rate of degradation out of proportion to the increased rate of synthesis. This increased catabolic rate is reflected by an increase in the activity of alkaline RNAase. As a consequence of this increased rate of RNA turnover, the normal polysome pattern is disrupted and there is a shift in the direction of smaller aggregates which are less efficient in synthesizing proteins. The net effect of malnutrition, therefore, is to decrease protein synthesis, but certain proteins, that is, RNA polymerase and alkaline RNAase, are spared and their synthesis is even increased.

These changes occurring at the cellular level with malnutrition have enabled us to "mark" certain tissues as being malnourished. This "marking" may prove particularly useful for diagnostic purposes when tissue is available during the period of malnutrition or shortly thereafter. This is particularly useful when one wishes to examine changes in human tissues, since certain of these changes are reflected in white cells, in plasma, in urine, and even in amniotic fluid, opening new vistas for diagnosis and monitoring of therapy, not only in postnatal malnutrition but also in "fetal malnutrition."

IV. HUMAN BRAIN GROWTH

The sequential changes in cellular growth of human brain are not as clearly defined as in the animal studies. In addition, there are presently no studies in the human relevant to the mechanisms involved in producing these changes. What is known, then, is strictly descriptive. We have previously demonstrated (12) that DNA content increases linearly in whole human brain during the prenatal period, begins to taper off around birth, and reaches a maximum at around 1 year of age. More recent data (13) have demonstrated two peaks of cell division in human brain, one around 26 weeks gestation and a second around birth. They have ascribed the first peak to neuronal division and the second to proliferation of glia. In two studies carried out at the Hospital Roberto del Rio in Santiago, Chile (14, 15), it was shown that severe malnutrition during the first year of life, resulting in the death of the infant, results in a reduced number of cells in the brain as well as a reduction in total lipid, cholesterol, and phospholipid content. If the malnutrition is prolonged into the second year, then the lipid/DNA ratio or lipid per cell is also reduced. The rate of cell division in various regions of the human brain and the effect of malnutrition on those regions has also been studied (16). The results indicate that the rate of cell division postnatally is about the same in cerebrum and cerebellum and that in both of these regions, as well as in brainstem, DNA synthesis stops at about the same time (approximately 1 year of age). Malnutrition reduces the number of cells in all three regions.

In addition to these studies, it has been demonstrated that head circumference is reduced in children who died of undernutrition during the first year of life and that this reduction reflects accurately the chemical changes just described (17).

Thus, the descriptive changes which have been demonstrated to occur in human brain are similar to those previously described in animal brain. Still unknown is whether or not the mechanisms producing these changes are similar, and what the functional implications of these changes are.

V. SUMMARY

Malnutrition during early brain development has been shown to curtail both protein synthesis and DNA synthesis. The result is a permanently stunted brain containing fewer cells of normal size. The same degree of malnutrition after cell division has stopped will retard net protein synthesis and result in a brain with a normal number of smaller cells. These changes are reversible with subsequent refeeding. Thus protein synthesis is curtailed and, if DNA synthesis is occurring, it too is curtailed. Our data suggest that the regulation of RNA metabolism and hence protein synthesis may be affected by malnutrition in two ways: elevation of alkaline RNAase activity and limitation of nucleotide pool size. In addition, during proliferative growth this limited pool size may retard DNA synthesis. Finally certain proteins, such as DNA polymerase, that are necessary for DNA synthesis may also be reduced during early malnutrition.

In the human brain DNA synthesis and myelination are also curtailed by malnutrition occurring during the time that these processes are most active. The mechanisms for these changes as well as their functional significance are as yet unknown.

ACKNOWLEDGMENT

This research was supported by National Foundation grant no. CRBS-262, National Institutes of Health grant no. HD-03888 and Nutrition Foundation grant no. 398.

REFERENCES

1. Boivin, A., Vendrely, R., and Vendrely, C. (1948): *Compt. Rend. Acad. Sci. (Paris)*, 226:1061.
2. Enesco, M., LeBlond, C. P. (1962): *J. Embryol. Exptl. Morphol.*, 10:530.
3. Winick, M., and Nobel, A. (1966): *J. Nutr.*, 89:300.
4. Fish, I., and Winick, M. (1969): *Pediat. Res.*, 3:407.
5. Fish, I., and Winick, M. (1969): *Exptl. Neurol.*, 25:534.
6. Winick, M. (1970): *Fed. Proc.*, 29:1510.
7. Culley, W. J. (1971): *Fed. Proc.*, 30:459.
8. Brasel, J. A., Ehrenkranz, R. A., and Winick, M. (1970): *Devel. Biol.*, 23:424.
9. Munro, H. N., and Drysdale, J. W. (1970): *Fed. Proc.*, 29:1469.
10. Miller, S. A. (1970): *Fed. Proc.*, 29:1497.
11. Metcoff, J. (*in press*): *Proceedings of Symposium on Nutrition and Fetal Development* sponsored by the National Foundation, New York, 1972.
12. Winick, M. (1968): *Pediat. Res.*, 2:350.
13. Dobbing, J., and Sands, J. (1970): *Nature*, 226:639.
14. Winick, M., and Rosso, P. (1969): *Pediat. Res.*, 3:181.
15. Winick, M., Rosso, P., and Waterlow, J. (1970): *Exptl. Neurol.*, 26:393.
16. Rosso, P., Hormazábal, J., and Winick, M. (1970): *Am. J. Clin. Nutr.*, 23:1275.
17. Winick, M., and Rosso, P. (1969): *J. Pediat.*, 74:774.

Dietary Lipids and Postnatal Development
Raven Press, New York © 1973

Dietary Lipids and Brain Development

C. Galli

Institute of Pharmacology, University of Milan, Milan, Italy

I. INTRODUCTION

In recent years considerable evidence has accumulated in various laboratories indicating that insufficient nutrition delays the maturation of the brain and results in reversible and/or irreversible damage to the central nervous system (CNS) from a structural, and possibly from a functional, point of view. In contrast, the chemical composition of the brain is considered to remain relatively constant after maturity, and to be more resistant than other organs to the influences of external factors. Most investigations have been concerned with studying the effects of calorie and/or protein reduction in the diet, whereas fewer experiments have been carried out to investigate the effects of prenatal and early postnatal lipid deficiency. Previous investigations have mainly studied the effects of essential fatty acid (EFA) deficiency on the CNS in weaning animals, that is, at a late stage in nervous development, and have observed rather limited alterations in brain fatty acid composition (1–5).

Over the past few years we have been studying the consequences on the developing brain of a nearly complete dietary fat deprivation, resulting in a deficiency of EFA, which are normally present in high concentrations in structural phospholipids of the CNS. Only traces of lipids were detected in the deficient diet, presumably in connection with the presence of lipid-soluble vitamins.

To induce the deficiency at an early stage of brain maturation, pregnant rats were fed either a semisynthetic control diet containing 2% corn oil or a fat-free diet, and this dietary treatment was continued with the young rats, after weaning at 30 days, for different periods of time. A similar experimental approach to study the CNS in EFA-deficient animals has been used by Berg-Hansen and Clausen (6) and by Sun (7), who have investigated the effects on brain lipid composition. Steinberg et al. (8), instead, fed weanling female rats a completely fat-free diet until they were mated, and measured the weight of the brain and its lipid composition in the neonates. More recently, Svennerholm et al. (9, 10) have, using the same scheme of animal treatment, described the effects of different levels of EFA in a diet having a constant linoleic:linolenic acid ratio and a constant percentage of fat on brain and body weight and brain fatty acid composition of weanlings.

In our investigations we have studied the effects of EFA deficiency on brain and body weight, on the composition of phospholipids, and on their fatty acids after extraction from brain and various brain subcellular structures. The effects of rehabilitation were also studied. Most of this work has been presented in previous papers (11–16), which also describe in detail the experimental conditions and methods employed.

II. EXPERIMENTAL CONDITIONS

The diets used were modified from Aaes-Jørgensen and Holman (17); the deficient diet was virtually lipid-free (0.1% w/w of the diet or 0.24% of the calories) and contained 0.004% linoleate and 0.001% linolenate, whereas the control diet contained 2% corn oil (4.9% of the calories) with 0.83% linoleate and 0.023% linolenate and no cholesterol. The experimental approach, using diets with constant percentages of fat and different levels of EFA, maintaining a constant linoleic:linolenic acid ratio, is considered (8, 9) to reduce the number of experimental variables. However, in these diets the proportion of unsaturated versus saturated fatty acids varies. Furthermore, the linoleic:linolenic acid ratio differs considerably in various edible fats and, hence, difficulties are encountered in establishing the physiological dietary ratio.

The stomach contents of suckling rats from mothers fed, starting from 1 week before delivery, either the control or the fat-free diet were also analyzed for linoleic and arachidonic acid content in total lipids and phospholipids.

Groups of pregnant rats were fed either the control or the deficient diet, and the treatment was continued in the young male rats after weaning. The animals were sacrificed at 3, 10, 30, 60, 90, 180, and 365 days of age. Four additional groups of rats on a combination dietary program [initial period on the deficient diet (D), followed by a period on the control diet (C)], were also studied (groups 10 D + 20 C, 30 D + 30 C, 10 D + 50 C, 90 D + 90 C) (13). Body and brain weights were measured, and brains from all groups were pooled for lipid analysis. The individual variability of data for brain lipids in animals of the same group was tested in the 60-day-old rats and found to be very small.

Further groups of control and deficient animals were sacrificed at 140 days, various brain subcellular fractions (myelin, mitochondria, microsomes, synaptosomes) were purified, and the fatty acid composition of ethanolamine phosphoglyceride (EPG) was analyzed (16). The procedures for extraction of lipids, analysis of phospholipids, purification of EPG, the major highly unsaturated brain phospholipid, which is more significantly modified by EFA deficiency (11), and analysis of the fatty acid composition have been described previously (11–16).

Finally, labeled linoleic and stearic acids, complexed with albumin (17), were injected into the carotid of control and deficient adult rats. The animals were sacrificed at 1, 4, 8, and 24 hr, lipids were extracted from brain, liver, and plasma, the radioactivity of the lipid extracts was measured, and the specific activities were calculated.

III. RESULTS AND DISCUSSION

A. Effects of EFA Deficiency on Milk Lipids

The effects of the dietary treatments on the levels of linoleic and arachidonic acids in lipids from the stomach contents of suckling rats at various ages is shown in Table 1. It appears that linoleic acid in total lipids is considerably reduced by the deficient diet and is 40, 25, 20, and 12%, respectively, in 1-, 3-, 7-, and 14-day-old deficient rats. Arachidonic acid decreases in the stomach contents of EFA-deficient rats at a slower rate. Linoleic acid in phospholipids is decreased only after 3 days in the deficient group, and to a smaller extent than in total lipids. Arachidonic acid levels in milk phospholipids are not modified by the lipid composition of the maternal diet under our experimental conditions. These results indicate that the dietary intake of EFA is considerably reduced from the beginning of lactation in sucklings from mothers fed an EFA-deficient diet. The persistence of constant levels

TABLE 1. *Linoleic and arachidonis acid levels (percentage of total fatty acids) in total lipids (TL) and phospholipids (PL) from the stomach contents of suckling rats at various ages*

	1 day				3 days				7 days				14 days			
	C		D		C		D		C		D		C		D	
	TL	PL	TL	PL	TL	PL	TL	PL	TL	PL	TL	PL	TL	PL	TL	PL
18:2	7.2	7.6	2.9	7.9	5.3	7.8	1.3	3.3	5.2	6.0	1.1	3.2	4.8	6.6	0.6	2.5
20:4	3.6	11.9	2.2	11.3	1.3	11.6	0.7	11.3	1.0	11.3	0.4	8.0	1.5	8.9	0.1	8.8

C = control, D = deficient.

of arachidonic acid in milk phospholipids (which are less than 1% of milk fat), however, may represent an important mechanism in the continuous supply of long-chain polyunsaturated fatty acids, even by mothers kept on EFA-deficient diets.

B. EFA Deficiency and Brain Development

The effects of the different diets on brain and body weights in developing rats are shown in Table 2. Brain weights are significantly lower in the deficient animals at 3 months of age (13), whereas body weights are decreased after 60 days. Decreased body and brain weights in 4-month-old male and female EFA-deficient rats have also been reported by Sun (7), whereas Svennerholm et al. (9, 10) did not find reduction of the brain weight in 30-day-old rats maintained on a low-EFA diet.

TABLE 2. *Body and brain weights of growing male rats fed control diet (C), EFA-deficient diet (D) and EFA-deficient diet followed by control diet (D + C)*

Group	No. in group	Body weight (g)	Percent difference from control	Brain weight (mg)	Percent difference from control
10 C	11	21 ± 0.4*		955 ± 14*	
10 D	9	18 ± 0.4^a	−14	858 ± 8^a	−10
30 C	4	46 ± 1.0		1388 ± 7	
30 D	4	45 ± 1.6	−2	1346 ± 30	−3
10 D + 20 C	4	56 ± 1.0^a	+22	1477 ± 19^c	+6
60 C	6	177 ± 9		1702 ± 35	
60 D	5	108 ± 6^a	−39	1655 ± 21	−3
10 D + 50 C	7	147 ± 8^c	−17	1680 ± 12	−1
30 D + 30 C	3	180 ± 13	+2	1642 ± 54	−4
90 C	4	375 ± 8		1911 ± 46	
90 D	3	128 ± 1^a	−66	1517 ± 39^b	−21
180 C	2	400 (396, 404)		1838 (1840, 1836)	
180 D	2	214 (208, 220)	−46	1726 (1671, 1782)	−6
90 D + 90 C	2	385 (378, 392)	−4	1955 (1937, 1973)	+6

Values indicated as follows are significantly different from control values at the *p* levels shown: $a = p$ 0.001; $b = p$ 0.01; $c = p$ 0.05.
*Values are means ± S.E.

Brain lipid concentrations were found to be lower in 6-month- and 1-year-old EFA-deficient rats (14), whereas the brain phospholipid composition was not found to be altered (11). Unchanged brain phospholipid composition in EFA-deficient rats has also been reported by Sun (7).

The percentages of the three major fatty acid families of brain EPG in control and EFA-deficient rats at various ages are shown in Table 3. In normal development, $n - 9$ acids (mostly monoenes with 18 and 20 carbon atoms) are reduced to about two-thirds after 1 year, and $n - 3$ acids (mainly 22:6) do not change significantly up to 6 months but decrease later. In EFA-deficient animals, $n - 9$ acids increase at 30 days, and, in contrast, $n - 6$ are decreased after 10 days, with respect to control values; $n - 3$ acids decrease only after 6 months (13). The unsaturation index, which indicates the average quantity of unsaturation present in the fatty acid mixture (13), remains unchanged in the deficient group, indicating that the phospholipid molecule maintains a constant level of unsaturation of its fatty acid mixture regardless of the amount of EFA present in the diet.

TABLE 3. *Brain ethanolamine phosphoglyceride fatty acid families as molar percentage of total fatty acids in control (C) and EFA-deficient (D) rats*

Days	(n - 9) acids		(n - 6) acids		(n - 3) acids		U.I.	
	C	D	C	D	C	D	C	D
3	11.2	11.8	30.8	29.2	14.8	16.5	233	238
10	11.8	10.6	30.2	27.9	14.7	16.3	225	225
30	23.5	29.4	28.2	24.1	12.9	14.0	218	223
60	32.1	38.8	24.3	21.0	10.3	9.7	201	191
90	34.8	41.8	24.7	19.5	10.2	12.2	201	201
180	41.1	52.4	19.6	14.4	13.7	7.2	202	181
365	38.4	49.0	17.3	11.8	9.8	5.3	167	163

The changes observed in the $n - 9$ and $n - 6$ fatty acid families consist mainly of a rise of $n - 9$ trienes (20:3 and 22:3) and a reduction in $n - 6$ tetraenes (20:4 and 22:4). Similar findings have been reported by Sun (7), whereas, in the study of Svennerholm (9, 10) the presence of 22:3 $n - 9$ in brain EPG from EFA-deficient animals was not detected. The data reported by Steinberg et al. (8) for the fatty acid composition of brain EPG in newborn EFA-deficient and control rats are quite unrealistic and difficult to interpret. The increase of trienes and the decrease of tetraenes induced in brain EPG by EFA deficiency result in an increase of the triene:tetraene ratio, which represents a biochemical index of EFA deficiency (17). This is due to the removal of inhibition in the conversion of oleate to polyunsaturated members of the $n - 9$ fatty acid family, namely trienes, as a consequence of the lack of linoleic and linolenic acids in the diet (18). The triene: tetraene ratio increases linearly at least up to 6 months of age in the deficient animals, indicating that continuous replacement of fatty acids occurs even in the adult brain (Fig. 1).

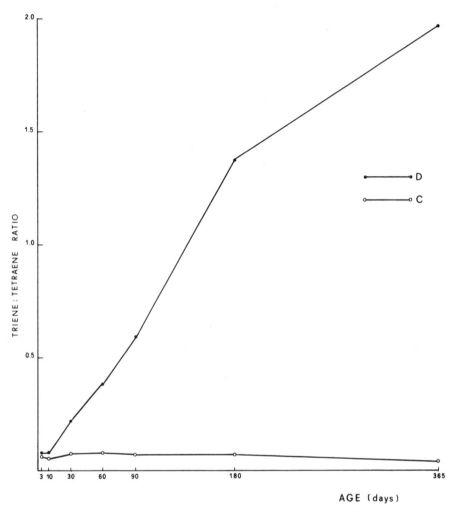

FIG. 1. Triene:tetraene ratios in brain ethanolamine phosphoglycerides of control (C) and EFA-deficient (D) rats at various ages.

C. Effects of EFA Deficiency on Brain Subcellular Fractions

Myelin, mitochondria, microsomes, and synaptosomes have been isolated from 140-day-old control and deficient male rats. Fatty acids of EPG have been analyzed, and the triene:tetraene ratios have been calculated in the various fractions. The values are presented in Table 4, which shows that the triene:tetraene ratio is more than double in myelin compared to the other

TABLE 4. *Triene:tetraene ratio in ethanolamine phosphoglyceride in whole brain and subcellular fractions*

Whole Brain		Synaptosomal fraction		Mitochondrial fraction		Microsomal fraction		Myelin fraction	
C	D	C	D	C	D	C	D	C	D
0.053	1.264	0.083	0.724	0.016	0.752	–	0.778	0.132	1.920

C = control; D = deficient.
From C. Galli, H. I. Trzeciak, and R. Paoletti (1972): *J. Neurochem.,* 19:1863. Reprinted with the permission of the publishers.

fractions. This indicates that the fatty acid composition of myelin lipids, which, however, are considered to be relatively stable, is more significantly modified than that of other structures and suggests continuous accumulation of trienes in myelin in EFA deficiency.

D. Reversal of EFA Deficiency in Growing Rats

Feeding the control diet to rats kept deficient for a period of time causes a reduction of n - 9 and an elevation of n - 6 acids with respect to control values, as shown in Table 5 (13). The tetraene members of the n - 6 family

TABLE 5. *Brain EPG fatty acid families in control (C) and EFA-deficient (D) rats and EFA-deficient rats put on control diets (D + C)*

	30 days			60 days				180 days		
	C (4)*	10 D + 20 C (4)	D (4)	C (6)	10 D + 50 C (7)	30 D + 30 C (3)	D (5)	C (2)	90 D + 90 C (2)	D (2)
n - 9 acids	23.5	20.3	29.4	32.1	30.8	30.2	38.8	41.1	33.9	52.4
n - 6 acids	28.2	30.5	24.1	24.3	26.8	25.8	21.0	19.6	23.8	14.4
n - 3 acids	12.9	15.3	14.0	10.3	10.5	10.6	9.7	13.7	14.4	7.2

*Number of animals in group.
Values are expressed as molar percentages of total EPG fatty acid.

in particular (not shown in this table) are consistently higher in the (D + C) group with respect to the control (C) group. This "rebound" phenomenon suggests increased uptake of EFA by the brain of EFA-deficient rats, as previously shown in the liver (19), and active passage of fatty acids, even in the adult brain, as shown by other investigators (20–23).

E. EFA Deficiency and Myelin Fatty Acids

The fatty acid composition of EPG purified from myelin isolated from the 180-day-old rat groups (180 C, 90 D + 90 C, 180 D) reveals changes in the deficient group similar to those reported for whole brain (Table 6), that is, elevation of trienes and reduction of tetraenes. Similar findings have been reported by Sun (7). These changes practically revert in the D + C group, indicating that the turnover of myelin lipids or the exchange of their fatty acids continues even in the adult brain, in contrast with the concept that myelin is a rather stable structure. The "rebound" phenomenon of n - 9 and n - 6 fatty acids, however, does not appear in myelin EPG of the D + C group.

TABLE 6. *Fatty acid distribution (molar percentage) of ethanolamine phosphoglyceride in myelin of 180-day-old male rats*

Fatty acids	6 months Control	3 months + 3 months Deficient + Control	6 months Deficient
Saturates	14.2	14.7	14.5
Monoenes	65.1	64.7	61.5
Polyenes	20.7	20.5	24.0
Dienes	1.8	1.9	1.8
Trienes	2.3	3.9	12.2
Tetraenes	12.9	11.9	6.8
Pentaenes	0.5	1.0	1.6
Hexaenes	3.2	1.8	1.6
$(n$ - 9) acids	64.9	67.0	72.7
$(n$ - 6) acids	16.4	14.8	9.7
$(n$ - 3) acids	3.4	1.9	1.8
Unsaturation index	149.0	144.0	147.0
Triene:tetraene	0.18	0.33	1.79

F. Uptake of Fatty Acids by the Brain of EFA-Deficient and Control Rats

The "rebound" phenomenon of brain n - 6 fatty acids in the deficient rats placed on the control diet prompted us to investigate the uptake of fatty acids by brain in EFA deficiency. We injected 8 μC of 1-^{14}C labeled linoleic acid, in the form of albumin complex (21), in the carotid of 6-month-old control and deficient rats. Two animals from each group were sacrificed

1, 4, and 24 hr after injection. Lipids were extracted from brain, liver, and plasma, and the specific activities measured. The relative activities (DPM/ mg lipid in tissue:DPM/mg lipid in plasma), indicating the distribution of radioactivity in the recovered lipid fraction between tissue and plasma at various intervals of time, have been determined. The values obtained (average of two determinations) are presented in Fig. 2, which shows that at each time interval the uptake of radioactivity in the brain and liver of the deficient animals is twice as high as in the corresponding control tissues. Increased uptake of EFA in tissues of deficient animals has already been described (19), together with reduced turnover of the formed polyunsaturated fatty acids (24). Intracarotid injections of 1-¹⁴C stearic and linoleic acids were also made in 18-month-old rats and the radioactivity measured in the lipid extracts from plasma, brain, and liver 8 hr later. The relative activities (brain/plasma and liver/plasma) are shown in Fig. 3. It appears that the values are higher for both stearic and linoleic acids in the deficient animals, indicating increased incorporation of both saturated and unsaturated fatty acids in tissue lipids in EFA deficiency. It can be noted that the

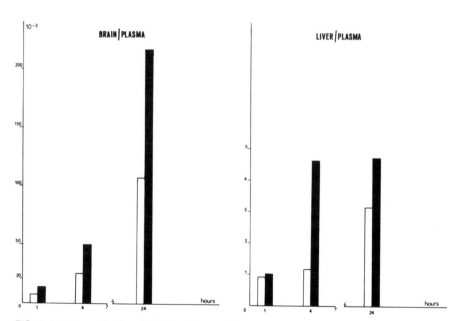

FIG. 2. Relative activities (DPM/mg tissue: DPM/mg plasma) of brain and liver total lipids after intracarotid injection of 1-¹⁴C-linoleic acid, in 6-month-old control (*white bars*) and EFA-deficient (*black bars*) rats.

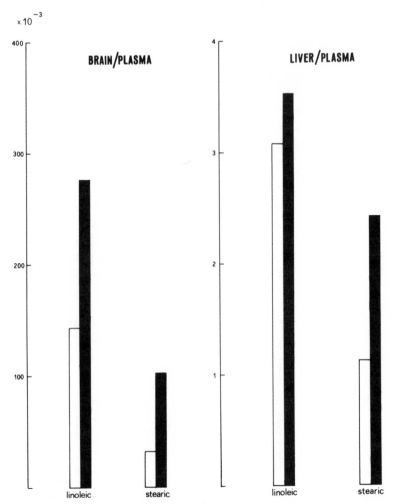

FIG. 3. Relative activities (DPM/mg tissue: DPM/mg plasma) of brain and liver total lipids 8 hr after intracarotid injection of 1-¹⁴C-linoleic and 1-¹⁴C-stearic acid in 18-month-old control (*white bars*) and EFA deficient (*black bars*) rats.

incorporation of fatty acids in the brain of both control and deficient rats is appreciable even in old animals. Previous investigators have reported active uptake of fatty acids by the brain of adult rats (20–22, 24). Preliminary data obtained in our laboratory suggest that the increased incorporation of stearic acid in the liver of deficient animals is associated with increased conversion to polyunsaturated acids of the n - 9 family.

IV. CONCLUSIONS

The results obtained in our investigation, in accordance with findings reported by other investigators, indicate that the developing CNS is affected by dietary deficiency of EFA. Although the developing brain is less sensitive than other organs to this deficiency, the effects of this treatment may be of comparable gravity with those obtained with other types of malnutrition (protein and/or calorie). These alterations, however, appear to be reversible.

More specifically, our data lead to the following general considerations: (a) The effects of EFA deficiency on the brain of growing rats consist mainly of a reduction of n - 6 and an increase of n - 9 triene fatty acids of brain phospholipids. These changes, which are monitored by an elevation of the triene:tetraene fatty acid ratio, progress steadily up to 6 months of age. (b) In spite of these considerable variations, the amount of unsaturation in each lipid class remains constant. The different stereochemical configuration of the newly formed n - 9 trienes substituting n - 6 tetraenes may not, however, have the same physiological significance. (c) The observed alterations differ in various brain subcellular structures and are more pronounced in myelin. (d) The fatty acid changes revert upon feeding the control diet after an equally long period on the deficient one. The recovery is nearly complete even in myelin. (e) Increased uptake of fatty acids is observed in the brain of adult deficient rats, and increased levels of EFA are found in the brain of the recovered animals.

These observations indicate that the fatty acids of brain lipids undergo continuous replacement. This process is affected by the levels of unsaturated fatty acids in the diet, as indicated also by studies showing accumulation of the long-chain polyunsaturated fatty acid 22:6 in the brain of rats fed a diet rich in n - 3 fatty acids (26). Changes in the fatty acids of membrane lipids are reported to alter membrane permeability, and activity of membrane-bound enzymes, mainly in the liver [see the review by Holman (25)]. Similar effects may be detectable in the CNS, and its functional activities may be impaired.

Behavioral studies also carried out in our investigations indicate that the performance of EFA-deficient animals is altered (15, 26). According to Caldwell and Churchill (27), the behavioral alterations induced by EFA deficiency are irreversible, in spite of the reversibility of the chemical changes (13). However, the number of variables involved in this type of investigation raises considerable difficulties in the interpretation of the data.

Further investigations are called for to explore, more broadly, brain alterations and their reversibility in EFA deficiency, both from a biochemical and a functional point of view. The possibility that dietary deficiencies of this type are present in human subjects in various parts of the world should also be carefully evaluated. This would be of great importance, from a practical point of view, in enabling the design of appropriate dietary formulas and nutritional conditions to ensure adequate intake of EFA in children during the suckling period.

REFERENCES

1. Witting, L. A., Harvey, C. C., Century, B., and Horwitt, M. K. (1961): *J. Lipid Res.*, 2:412.
2. Mohrhauer, H., and Holman, R. T. (1963): *J. Neurochem.*, 10:523.
3. Biran, L. A., Bartley, W., Carter, C. S., and Renshaw, A. (1964): *Biochem. J.*, 93:492.
4. Rathbone, L. (1965): *Biochem. J.*, 97:629.
5. Walker, B. L. (1967): *Lipids*, 2:497.
6. Berg-Hansen, J., and Clausen, J. (1969): *Z. Ernährungswiss.*, 9:278.
7. Sun, G. Y. (1972): *J. Lipid Res.*, 13:56.
8. Steinberg, A. B., Clarke, G. B., and Ramwell, P. W. (1968): *Develop. Psychobiol.*, 1:225.
9. Alling, C., Bruce, Å., Karlsson, I., Sapia, O., and Svennerholm, L. (1972): *J. Nutr.*, 102:773.
10. Svennerholm, L., Alling, C., Bruce, Å., Karlsson, I., and Sapia, O. (1972): In: *Lipids, Malnutrition and the Developing Brain*, A Ciba Foundation Symposium, edited by K. Elliott and J. Knight. Elsevier, Excerpta Medica and North Holland, Amsterdam.
11. Galli, C., White, H. B., Jr., and Paoletti, R. (1970): *J. Neurochem.*, 17:347.
12. Galli, C., White, H. B., Jr., and Paoletti, R. (1971): In: *Chemistry and Brain Development*, edited by R. Paoletti and A. N. Davison. Plenum Press, New York.
13. White, H. B., Jr., Galli, C., and Paoletti, R. (1971): *J. Neurochem.*, 18:869.
14. Galli, C., White, H. B., Jr., and Paoletti, R. (1971): *Lipids*, 6:378.
15. Paoletti, R., and Galli, C. (1972): In: *Lipids, Malnutrition and the Developing Brain*, A Ciba Foundation Symposium, edited by K. Elliott and J. Knight. Elsevier, Excerpta Medica and North Holland, Amsterdam.
16. Galli, C., Trzeciak, H. I., and Paoletti, R. (1972): *J. Neurochem.*, 19:1863.
17. Holman, R. T. (1960): *J. Nutr.*, 70:405.
18. Holman, R. T., and Mohrhauer, H. (1963): *Acta Chim. Scand.*, 17:584.
19. Stein, O., and Stein, Y. (1964): *Biochim. Biophys. Acta*, 84:621.
20. Dhopeshwarkar, G. A., and Mead, J. F. (1969): *Biochim. Biophys. Acta*, 187:461.
21. Dhopeshwarkar, G. A., and Mead, J. F. (1970): *Biochim. Biophys. Acta*, 210:250.
22. Dhopeshwarkar, G. A., Subramanian, C., and Mead, J. F. (1971): *Biochim. Biophys. Acta*, 231:8.
23. Oldendorf, W. H. (1971): *Proc. Soc. Exp. Biol. Med.*, 136:385.
24. Harris, P. M., Robinson, D. S., and Getz, G. (1960): *Nature*, 188:742.
25. Holman, R. T. (1968): In: *Progress in the Chemistry of Fats and Other Lipids*, Vol. IX, edited by R. T. Holman. Pergamon Press, Elmsford, N.Y.
26. Galli, C., Trzeciak, H. I., and Paoletti, R. (1971): *Biochim. Biophys. Acta*, 248:449.
27. Caldwell, D. F., and Churchill, J. A. (1966): *Psychol. Rep.*, 19:99.

Dietary Lipids and Postnatal Development
Raven Press, New York © 1973

The Effect of Dietary Lipids on the Central Nervous System

C. Alling, Å. Bruce, I. Karlsson, and L. Svennerholm

Department of Neurochemistry, Psychiatric Research Centre, University of Göteborg, Göteborg, Sweden

I. INTRODUCTION

Of the various dietary fatty acids, linoleic and linolenic acids have received most attention in studies of the effects of dietary lipids on the central nervous system. This is natural because they cannot be synthesized by mammals but must be supplied with the diet. They are precursors of the two series of polyunsaturated fatty acids which constitute a large portion of the fatty acids in brain. They are traditionally considered essential, but it has not been definitively proved that linolenic acid really is an essential fatty acid (1, 2). It is, however, necessary to study both the linoleic and the linolenic acid series simultaneously, because the tissue concentrations of fatty acids of one of the families are dependent on the concentrations of the fatty acids of the other (3, 4, 5, 6).

The present report concerns the effects of three dietary levels of essential fatty acid (EFA) on the lipid and fatty acid composition of cerebrum during prenatal and postnatal development in the rat. A few studies have previously been undertaken in which the lipid and fatty acid composition was determined in brain of the offspring of rats fed an EFA-free diet from about 1 week before delivery. Such a feeding procedure will not result in EFA deficiency until after the suckling period (7, 8, 9). In all our studies the female rats were fed a low-EFA diet before mating, as were the offspring during the whole gestation period, lactation, and later postnatal development.

II. MATERIAL AND METHODS

Rats of the Sprague-Dawley strain were fed *ad libitum* synthetic diets containing three different levels of EFA, 3.00, 0.75, and 0.14 cal-%, respectively. The caloric compositions of the diets were protein 16, carbohydrates 63, and fat 21 cal-%. A detailed description of the diets is given in Table 1. The rats were fed the experimental diets with 3 and 0.75% cal-% of EFA for more than two generations before mating for the actual experiment. Rats on the 0.75% EFA diet were fed the 0.14% EFA diet from 3 weeks before mating. The rats were kept in rooms with controlled duration of daylight (12 hr), temperature (23°C), and humidity (60%). In our previous studies no such air-controlled room was available, which resulted in an increased mortality of the offspring when the concentration of EFA in the diet of the female was reduced to less than 1 cal-%. The dietary intake of the female was, therefore, not reduced below 0.75 cal-%. By careful control of the environmental factors, we were able to keep the offspring alive on a diet containing hydrogenated lard as the only fat source. The fat was assumed to be completely hydrogenated, but gas-liquid chromatography revealed 3.6% monoenoic fatty acids and 0.6% and 0.2% of two peaks with the same equivalent chain length as linoleic and linolenic acid, respectively. No attempt was made to prove that the two peaks corresponded to the actual fatty acids. It might also be possible that they were, at least partly, *trans-trans* acids without any biological effect (10).

The females were mated at the age of 90 days. On the day of birth each litter was reduced to six offspring. The litters were weighed every second

TABLE 1. *Composition of the diets*

Cal-% EFA	3.00	0.75	0.14
	Constituents in 1 kg of diet (g)		
Protein (fish protein EFP 90)	178	178	178
Corn starch	620	620	620
Sucrose	52	52	52
Fat mixture	100	100	100
Hydrogenated lard	78.0	94.5	100
Sunflower oil	16.5	4.1	—
Linseed oil	5.5	1.4	—
Salt mixture[a]	40	40	40
Vitamin mixture[b]	2.0	2.0	2.0
Choline chloride	2.0	2.0	2.0
Cellulose	36	36	36

[a]USP 17 + per kg of salt 0.088 g $KAl(SO_4)_2$, $12H_2O$; 0.28 g NaF; 0.009 g $NaAsO_2$; 0.022 g $Na_2B_4O_7$, $10H_2O$, and 0.0031 g Na_2MoO_4, $2H_2O$.

[b]Vitamins in 1 kg of diet: retinol, 1000 IU; ergocalciferol, 500 IU; (mg) thiamin, 50; riboflavin, 20; pyridoxine, 20; nicotinamide, 200; pantothenic acid, 100; p-aminobenzoic acid, 100; menaquinone, 10; biotin, 1; folic acid, 5; cyanocobalamin, 0.005; myo-inositol, 1000; and tocopheryl acetate, 500.

day after the fifth day and every fifth day after the 21st day. Only animals with a weight within 2 S.D. of the mean weight were used for biochemical analysis. In experiments on animals at most 18 days old, sex was ignored. At the age of 30 days, equal numbers of females and males were used for the determinations, and at higher ages only males were used. The blood was obtained by bleeding the animals after decapitation. The brain was cut through the olfactory lobes and the cerebral peduncles, and only the cerebrum was used for analysis. The brains and serum from each litter were pooled and stored at −20°C until analyzed.

On the fourth and 14th day of lactation the litters were taken away from their mothers, and 4 hr later the females were given 4 IU of oxytocin. After 15 min they were anesthetized with ethyl ether. The milk (0.2 to 1.0 ml) was collected by intermittent suction.

The brain was homogenized at +4°C in a Potter-Elvehjelm all-glass homogenizer. The lipids were extracted from brain homogenate, blood serum, and milk with chloroform-methanol, 1:1, and the extracts were freed from nonlipid material by solvent partition (11, 12).

Portions of the lipid extract were analyzed for major lipid classes with methods used in our laboratory for several years and particularly adapted for the nutritional experiments (11, 13). Portions of the lipid extracts were extracted by thin-layer chromatography (TLC). The plates were developed with chloroform-methanol-water, 65:25:4, for the isolation of phospholipids, and with light petroleum-diethyl ether-acetic acid, 85:15:1, for the isolation of triglycerides. Their fatty acid composition was determined by gas-liquid chromatography of their methyl esters on a DEGS (diethyleneglycolsuccinate polyester) column (11).

III. BODY GROWTH AND BRAIN WEIGHT

The growth curves (Fig. 1) from term up to 120 days of age were studied in five to eight litters in each of the three dietary groups. Only the weights up to 90 days are given, since some litters became infected after 90 days of age. In all age groups the mean body weight of the 3% EFA group was slightly higher than in the 0.75% EFA group. The weights of the two dietary groups between the 25th and 60th day were analyzed statistically. The difference in weight was significant both for male rats ($p < 0.005$) and female rats ($p < 0.0095$). The weight curve for the 0.14% EFA group was distinctly lower than that for the 0.75% EFA group (Fig. 1). Statistical analysis of the weights between the 25th and 60th day showed that the difference was significantly smaller at the 0.0005 level both for male and female rats in the 0.14% EFA group than in the 3.0 or 0.75% EFA group.

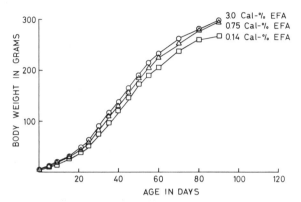

FIG. 1. Growth curves for male rats fed three dietary levels of EFA up to the age of 90 days.

It has long been known that EFA deficiency affects body growth. When Burr and Burr (14) excluded fat from the diet of rats, the animals showed retardation of body growth. The research group, headed by Deuel, later found that growth was directly related to the intake of linoleate in rats fed an otherwise identical diet (15). A reduced growth rate of rats fed diets deficient in EFA has been used by subsequent investigators as a criterion of EFA deficiency (16, 17, 18, 19). The rats have in general been made deficient by excluding fat from the diet starting at weaning, but withdrawal of dietary fat leads to a significant change in the fatty acid metabolism (20). The reduced body weight might thus depend at least partly on the diminished fat intake. However, Sinclair and Collins (18, 21), who used the same nutrient composition except for variation of the levels of EFA in experimental and control diets, observed a significantly reduced growth rate in the EFA-deficient rats.

In the previous studies, referred to above, the drastic reduction of the dietary EFA intake was introduced in association with weaning or later. Studies of the protein metabolism of protein-deficient animals have shown (22) that if the deficiency is induced early enough, the animals will adapt themselves to these conditions and will be able to minimize the effects of the deficiency. We assume that such an adaptation can also occur to EFA deficiency. However, it is unlikely that a metabolic adaptation to a low EFA intake can occur to the same degree after weaning as when the fetus has been receiving a low EFA supply from the beginning of gestation. Even when the females and the offspring were adapted to a low EFA intake throughout the period of gestation and lactation, the growth rate of the rats was diminished.

In a recent study we found that the brain weight in 21-day-old rats that had received 0.75% EFA did not differ from that in rats that had received 5% (13). In none of the age groups in the present study was any difference in brain weight found between the three dietary groups.

Paoletti and co-workers (8, 9) reported diminished body and brain weights of EFA-deficient rats at an early age, even at 10 days, although the EFA reduction was started 5 days before term. The curves for body and brain weights during growth in their studies also showed a wide range of variation in the controls, and we do not consider that the animals studied at the different ages were representative. When the brain weights were related to the body weight, no significant differences were found between the control and experimental groups.

IV. MILK COMPOSITION

We have been unable to determine the maternal supply of EFA to the offspring during gestation and lactation. But in order to get a rough estimate, we determined the concentration of EFA in the milk on the fourth and 14th day of lactation. Since it was technically difficult to analyze the fatty acid composition of the total milk lipids, and since 96 to 99% of the milk fatty acids were derived from the triglycerides, only this lipid fraction was analyzed by gas-liquid chromatography.

The triglyceride concentration of the milk, determined with an enzymatic assay for glycerol, was lowest both on the fourth and the 14th day in the 3% EFA group, where it was 150 and 120 mmoles/liter, respectively. The concentration was somewhat higher on the fourth day and the 14th day in

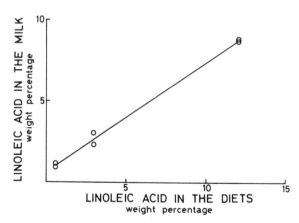

FIG. 2. Correlation between linoleic acid concentration in the triglycerides of the diets and in the triglycerides of the rat milk.

the 0.75% EFA group, 215 and 215 mmoles/liter, and in the 0.14% EFA group, 195 and 210 mmoles/liter.

Our major aim, however, was to find out whether the concentration of linoleic acid in the maternal milk varied with that in the diet. The results are given in Fig. 2. A direct linear correlation was found between the linoleate concentration in the diet and that of the milk triglycerides. The correlation was high, 0.99.

This correlation between the concentration of linoleic acid in the diet and the milk is a novel observation, but a certain relation has been demonstrated in the cow, sow, and mouse. In humans the concentration of linoleic acid in the milk triglycerides was shown to increase twofold during a 12-day period in which the ordinary dietary fat was replaced by sunflower oil (23).

V. FATTY ACID COMPOSITION OF SERUM LECITHINS

In order to estimate to what extent the rats suffered from EFA deficiency, we analyzed the fatty acid composition of serum lecithin.

Substantial differences in fatty acid patterns were found among the three dietary groups. Up to the age of 45 days the concentration of the sum of the linoleic and linolenic acid series in the 0.14 cal-% group was less than half of that in the 3.0 cal-% group. The 0.75 cal-% group had intermediate values. Newborns from the 0.75 and 0.14 cal-% groups had much lower values for the fatty acids of linoleic and linolenic acid series than the 3.0 cal-% group, indicating a large difference among the groups in the supply of EFA to the fetus. During the postnatal period the differences in concentration of the linoleic and linolenic acid series between the 0.75 and 3.0 cal-% groups were largest at 30 and 45 days of age, and between the 0.14 and 3.0 cal-% groups at 18 days of age. The smallest difference among the three groups was found at 120 days of age. The monoenoic acids (16:1 and 18:1) increased at the same time as the linoleic and linolenic acid series decreased. The principal developmental pattern showed no difference in the ratio between 18:2 (n - 6) and 20:4 (n - 6) among the groups.

Since the demonstration by Fulco and Mead (24) that, in the absence of EFA, eicosatrienoic acid, 20:3 (n - 9) is synthethized from oleic acid, the concentration of 20:3 (n - 9) in a lipid fraction has been used as a measure of EFA deficiency. As shown in Fig. 3, the concentrations of 20:3 (n - 9) in the 0.75 cal-% group were 2 to 5 molar-% and higher at all ages than in the 3.0 cal-% group. In the 0.14 cal-% group the concentrations of this fatty acid were about three times as high as in the 0.75 cal-% group. The maximum value for 20:3 (n - 9) was found at the same age, 30 days, as the

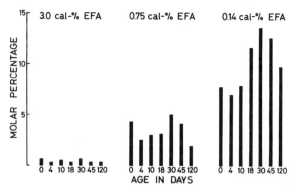

FIG. 3. Concentration of 20:3 (*n* - 9) in choline phosphoglycerides of serum from rats fed three dietary levels of EFA at different ages.

lowest value for the sum of the linoleic and linolenic acid series in the 0.75 and 0.14 cal-% groups.

These results have thus amply documented a biochemical deficiency of EFA in the 0.75 and 0.14 cal-% groups at birth and during the postnatal period.

VI. CONCENTRATIONS OF CEREBRAL LIPIDS

Table 2 gives the concentration of cholesterol, lipid-P, and fatty acids of phosphoglycerides, expressed as micromoles per gram fresh weight, from birth to 120 days of age, in the three dietary groups. The concentration of cholesterol increased about four times from birth up to the age of 45 days in the three groups. The increase in cholesterol concentration between 45 and 120 days of age was small. The concentration of lipid-P increased from birth up to the age of 30 days, after which the values were almost the same, and the concentration of fatty acids of phosphoglycerides increased in a similar way. There was no difference among the groups. Previous studies of phospholipid and cholesterol concentrations of rat brain during development could not be compared directly with those of the present study, since the whole brain was used for the analyses. In order to get a more accurate definition of the tissue material, only cerebrum was analyzed in this study. In another study from this laboratory (25), in which the same technique was used for isolating cerebrum, the phospholipid and ganglioside concentrations were determined from birth up to 21 days of age in offsprings from rats fed an ordinary pellet diet. The values for lipid-P found in the present study on the fourth, tenth, and 18th day of age were the same as those obtained in

TABLE 2. *Lipid composition of brain (μmoles/g fresh weight)*

Age in days	0	4	10	18	30	45	120
				3.0% EFA group			
Cholesterol	10	11	16	25	34	41	44
Lipid-P	29	29	36	50	68	67	66
Fatty acids of phosphoglycerides	49	52	60	82	97	100	104
				0.75% EFA group			
Cholesterol	12	9	16	27	38	42	44
Lipid-P	28	24	34	52	70	67	66
Fatty acids of phosphoglycerides	51	41	57	78	102	97	102
				0.14% EFA group			
Cholesterol	11	13	17	28	36	36	49
Lipid-P	25	29	35	50	68	66	67
Fatty acids of phosphoglycerides	45	50	62	91	102	96	95

that study at these ages. Galli et al. (8) have claimed that brain phospholipid concentrations are reduced after 90 days of EFA deficiency during postnatal development. However, their values for brain weight and lipid-P showed a great variance both for control and EFA-deficient animals, and, in our opinion, their results do not warrant such an interpretation. In our long-term experiment it was found that the concentrations of phospholipids were not influenced by the dietary concentration of EFA.

VII. FATTY ACID COMPOSITION OF CEREBRAL PHOSPHOGLYCERIDES

Gas-liquid chromatography analyses of the fatty acid composition of the total phosphoglyceride fraction revealed not only expected changes related to age (11, 26, 27), but also differences between age-matched animals in the three dietary groups. The choline and ethanolamine phosphoglyceride fractions were analyzed separately. The changes due to differences in dietary intake of EFA were most pronounced in the ethanolamine phosphoglycerides; the following presentation is therefore concerned only with the fatty acid composition of this phosphoglyceride.

Figure 4 shows the molar percentage of the sum of the fatty acid belonging to the linoleic and linolenic acid series. In all groups this sum changed, although only slightly. The highest amount was found at 10 and 18 days of age, after which it decreased. This decrease ran parallel to the increase

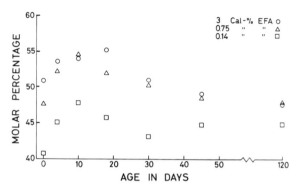

FIG. 4. Concentration of EFA in ethanolamine phosphoglycerides of cerebrum from rats fed three dietary levels of EFA at different ages.

of monoenoic acids, characteristic of myelin (Table 3). The difference between animals fed 3.0 and those fed 0.75 cal-% EFA was small. In all age groups except one, rats from the 0.75% EFA group had a somewhat lower value. The difference was largest in newborn rats. At that age the difference was due to a lower concentration of docosahexaenoic acid, 22:6 (n - 3), in the 0.14% EFA group, while the concentrations of fatty acids belonging to the linoleic acid series showed no such decrease. The concentrations of fatty acids of the linoleic and linolenic acid series of ethanolamine phosphoglycerides from rats fed 0.14 cal-% EFA were significantly lower than in the 3.0% EFA group at all ages. The differences were approximately the same during the suckling period, but later decreased and were found to be small at 120 days of age. The differences between the 0.14 and 3.0% EFA groups were due to pronounced reduction of docosahexaenoic acid, 22:6 (n - 3), in the 0.14% EFA group. The decrease in 22:6 (n - 3) was accompanied by an increase in 22:5 (n - 6) (Table 3). The concentration of 22:5 (n - 6) was several times higher in the 0.14% EFA group than in the 3.0% EFA group. However, since 22:5 (n - 6) was a minor fatty acid in both groups, the higher values in the 0.14% EFA group were not sufficient to maintain a concentration of EFA of the same magnitude as in the 3.0% EFA group. This reduction of 22:6 (n - 3) and increase of 22:5 (n - 6) might be due to the absolute amount of linolenic acid supply being insufficient *per se* or to an insufficient supply combined with a change in the ratio between linoleic and linolenic acids in the diet. A change in this ratio in the diet is known to influence greatly the concentration of 22:6 (n - 3) and 22:5 (n - 6) of tissue lipids, including brain (3, 4, 5, 6).

In addition to the differences in the concentrations of EFA of ethanolamine phosphoglycerides, the concentration of the fatty acid 20:3 (n - 9)

TABLE 3. Concentrations of major fatty acids of ethanolamine phospho-glycerides in cerebrum (molar-% of total fatty acids)

Age in days	0	4	10	18	30	45	120
			3.0% EFA group				
16:0	14	13	11	8	8	8	7
18:0	24	24	26	25	27	28	24
18:1	10	8	8	10	14	14	18
20:4 (n - 6)	18	21	23	22	18	16	16
22:4 (n - 6)	4	5	5	6	6	5	6
22:5 (n - 6)	7	4	3	2	2	1	1
22:6 (n - 3)	21	22	22	24	24	26	23
			0.75% EFA group				
16:0	15	14	11	9	8	7	7
18:0	23	23	25	28	27	26	25
18:1	11	9	8	9	13	16	17
20:4 (n - 6)	18	20	23	20	17	17	16
22:4 (n - 6)	4	5	5	5	5	5	5
22:5 (n - 6)	7	6	5	4	3	3	1
22:6 (n - 3)	18	20	21	23	24	23	25
			0.14% EFA group				
16:0	14	14	13	10	9	8	8
18:0	23	25	26	26	29	27	28
18:1	14	11	9	11	15	15	15
20:4 (n - 6)	19	20	22	20	17	16	14
22:4 (n - 6)	4	4	3	4	4	4	4
22:5 (n - 6)	7	8	7	7	9	9	7
22:6 (n - 3)	9	12	14	12	14	16	19

differed among the groups (Fig. 5). The concentrations were significantly increased in the 0.75% EFA group at all ages. However, a much more pronounced increase was found for the 0.14% EFA group. Of all ages, newborn rats had the highest value of 20:3 (n - 9) in the 0.75% EFA group, and in the 0.14% EFA group this age had the next highest value. During the first four days of extrauterine life the phospholipid accretion of rat brain was slow (25), and on the fourth day the concentration of 20:3 (n - 9) had fallen. After the fourth day the phospholipid accretion of brain increased, resulting in a greater requirement of EFA, and the concentration of 20:3 (n - 9) rose and, judging from the present results, reached a maximum at 18 days of age. The lowest value was found at 120 days of age, both in the 0.75 and 0.14% EFA groups. In summary, the results presented suggest that the concentration of 20:3 (n - 9) in brain ethanolamine phosphoglycerides depends on both the dietary supply of EFA and the phospholipid accretion in the brain.

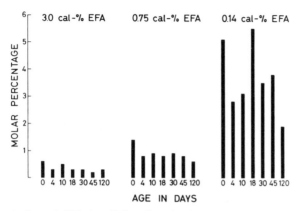

FIG. 5. Concentration of 20:3 (*n* - 9) in ethanolamine phosphoglycerides of cerebrum from rats fed three dietary levels of EFA at different ages.

In a study by Steinberg et al. (28), weaning female rats were given a fat-free diet and were mated 2 months later. Large alterations in the fatty acid composition of the brain lipids of the newborn offspring were reported, but the incompatible figures given for ethanolamine phosphoglycerides and total lipids reduce the value of the findings. Hansen and Clausen (7) fed pregnant rats diets free from EFA from 1 to 2 weeks before delivery, and analyzed the fatty acid composition of serum and brain in the offspring up to the age of 12 weeks. 20:3 (*n* - 9) in serum total lipids did not increase until 4 to 6 weeks postnatally, and no increase in 20:3 (*n* - 9) in brain was found. Paoletti and co-workers (8, 9) started to give rat dams an EFA-deficient diet, which was almost free from fat, 5 days before delivery. The fatty acid patterns of brain ethanolamine phosphoglycerides of the offspring at varying ages were analyzed. At 3 and 10 days of age no difference in the concentration of 20:3 (*n* - 9) was found between EFA-deficient rats and controls. At 30 days of age the concentration of 20:3 (*n* - 9) was slightly increased, and the largest differences were found in the oldest rats, 180 days of age. It is obvious that in these previous studies (7, 8, 9) the offspring were supplied during both the fetal and suckling period with considerable amounts of EFA by a release from maternal EFA depots. In the present investigation, however, the dietary supply of EFA was reduced in an earlier generation, so that the dams were depleted of EFA before mating. This enabled us to study the effects of a low EFA content of the maternal diet on the offspring during prenatal and postnatal life. Our results appear to warrant the conclusion that the cerebrum is less sensitive to long periods of low dietary supplies of EFA than other organs. When the dietary EFA reduction was moderate, only very small changes were observed in the brain ethanolamine phosphoglyceride fatty acid patterns. Not until we gave a very low dietary supply of EFA did apparent changes occur. These changes were most pronounced at 18 days of age, when the fatty acid pattern was influenced by the previous phase of rapid growth of the brain, and decreased when the animals became older and brain growth had ceased.

ACKNOWLEDGMENTS

We are indebted to Dr. H. Wendeus, Astra Nutrition, Mölndal, Sweden, for the gift of fish protein EFP-90, to Dr. R. Ohlsson, Karlshamns Oljefa-briker, Karlshamn, Sweden, for the hydrogenated lard, and to Dr. L. Wedberg, AB Ferrosan, Malmö, Sweden, for the vitamins. We thank Mrs. Karin Andersson and Mrs. Kristina Rinäs for their technical assist-ance, and Mr. Gösta Lundin for the care of the experimental animals. This work was supported by grants from the Swedish Medical Research Council (Project no. 13X–627) and Stiftelsen Svensk Näringsforskning.

REFERENCES

1. Holman, R. T. (1970): In: *Progress in the Chemistry of Fats and Other Lipids,* Vol. IX, part 5, edited by R. T. Holman. Pergamon Press, Oxford, p. 611.
2. Tinoco, J., Williams, M. A., Hincenbergs, I., and Lyman, R. L. (1971): *J. Nutr.,* 101:937.
3. Mohrhauer, H., and Holman, R. T. (1963): *J. Neurochem.,* 10:523.
4. Walker, B. L. (1967): *Lipids,* 2:497.
5. Galli, C., Trzeciak, H. I., and Paoletti, R. (1971): *Biochim. Biophys. Acta,* 248:449.
6. Svennerholm, L., Alling, C., Bruce, Å., and Sapia, O. (1972): In: *Lipids, Malnutrition and the Developing Brain,* A Ciba Foundation Symposium. ASP, Amsterdam, London, New York, p. 141.
7. Berg Hansen, I., and Clausen, J. (1968): *Z. Ernährungswiss.,* 9:278.
8. Galli, C., White, H. B., Jr., and Paoletti, R. (1971): *Lipids,* 6:378.
9. White, H. B., Jr., Galli, C., and Paoletti, R. (1971): *J. Neurochem.,* 18:869.
10. Holman, R. T. (1968): In: *Progress in the Chemistry of Fats and Other Lipids,* Vol. IX, part 2, edited by R. T. Holman. Pergamon Press, Oxford, p. 279.
11. Svennerholm, L. (1968): *J. Lipid Res.,* 9:570.
12. Olegård, R., and Svennerholm, L. (1970): *Acta Paediat. Scand.,* 59:637.
13. Alling, C., Bruce, Å., Karlsson, I., Sapia, O., and Svennerholm, L. (1972): *J. Nutr.,* 102:773.
14. Burr, G. O., and Burr, M. M. (1929): *J. Biol. Chem.,* 82:345.
15. Greenberg, S. M., Carbert, C. E., Savage, E. E., and Deuel, H. J., Jr. (1959): *J. Nutr.,* 41:473.
16. Panos, T. C., and Finerty, J. C. (1954): *J. Nutr.,* 54:315.
17. Mohrhauer, H., and Holman, R. T. (1963): *J. Lipid Res.,* 4:151.
18. Sinclair, A. J., and Collins, F. D. (1968): *Biochim. Biophys. Acta,* 152:498.
19. Aaes-Jörgensen, E., and Hölmer, G. (1969): *Lipids,* 4:501.
20. Smith, S., and Abraham, S. (1970): *Arch. Biochem. Biophys.,* 136:112.
21. Sinclair, A. J., and Collins, F. D. (1970): *Brit. J. Nutr.,* 24:971.

22. Waterlow, J. C. (1968): In: *Calorie Deficiencies and Protein Deficiencies,* edited by R. A. McCance and E. M. Widdowson. J. A. Churchill, London, p. 61.
23. Krámer, M., Szöke, K., Lindner, K., and Tarján, R. (1965): *Nutr. Dieta,* 7:71.
24. Fulco, A. J., and Mead, J. F. (1959): *J. Biol. Chem.,* 234:1411.
25. Vanier, M. T., Holm, M., Öhman, R., and Svennerholm, L. (1971): *J. Neurochem.,* 18:581.
26. Marshall, E., Fumagalli, R., Niemiro, R., and Paoletti, R. (1966): *J. Neurochem.,* 13:857.
27. Sinclair, A. J., and Crawford, M. A. (1972): *J. Neurochem.,* 19:1753.
28. Steinberg, A. B., Clarke, G. B., and Ramwell, P. W. (1968): *Develop. Psychobiol.,* 1:225.

Effects of Dietary Lipids on Lipid Biosynthesis and Exchange Reactions in Brain

Giuseppe Porcellati

Department of Biochemistry, The Medical School, University of Perugia, Policlinico Monteluce, Perugia, Italy

I. INTRODUCTION

It is known that the rate of *de novo* synthesis of phospholipids in animal tissue is probably controlled at the level of the cytidylyltransferase reaction (1–6). 1,2-Diacyl-*sn*-glycerophosphorylcholine (GPC) and monoacyl-*sn*-GPC (lysolecithin) are able to stimulate *in vitro* the cytidylyltransferase reaction of hepatic and nervous tissues (1, 3, 6, 7) and the fatty acyl profiles of the stimulating phospholipid molecules modify to a different extent the activating effects, probably modulating the regulatory action at the level of the rate-limiting cytidylyltransferase stage (3, 7, 8).

The fatty acyl composition of the membrane lipids is also of considerable interest in exerting regulatory effects on the base-exchange reaction, an interesting type of enzymic mechanism for the synthesis of phospholipid molecules at the expense of endogenous lipid (7, 9). The fully saturated species of 1,2-diacyl-*sn*-glycerophosphorylethanolamine (GPE) were found to exchange their base less readily than the unsaturated compounds, whereas the mixed species were more active in this respect (7, 9).

Studies carried out on essential fatty acid deficiency in rats, on the other hand, have shown that a dietary fat deficiency initiated in pregnant rats well before delivery and continued in the newborns for different intervals of time causes noticeable alterations in the fatty acid composition of brain phospholipids in the litters (10, 11).

Experiments have therefore been carried out with the aim of demonstrating whether an essential fatty acid (EFA) deficiency produced in rats can affect *in vitro* some enzymic activities involved in phospholipid biosynthesis (6) or in base-exchange reactions (12), which occur predominantly in membranes (5, 6, 13). Some activities related to choline phosphoglyceride

(CPG) synthesis in brain have been shown to be affected by the dietary treatment, whereas the formation of ethanolamine phosphoglyceride (EPG) is not influenced. Base-exchange reactions are slightly affected.

II. EXPERIMENTAL PART

All the experiments performed in this investigation were carried out on 6-month-old female Wistar rats raised on a fatty acid-deficient diet (General Biochemicals, Chagrin Falls, Ohio) as described by Sun (11). The treatment was initiated in pregnant rats 10 days before delivery and continued for 6 months. The controls were given the fatty acid-deficient diet supplemented with corn oil (5%, w/w) for the same period. Both the treated and the control animals were individually caged and supplied with water and food *ad libitum.* They were rigorously kept under the same hygienic, ambiental, and thermic conditions, and were killed by decapitation at the given intervals. The effects of the EFA deficiency were followed and checked systematically (11, 14). Only animals showing signs of fatty acid deficiency were used, but these constituted the majority.

Brain microsomes were prepared and purified as previously reported (5, 15). Pellets were washed carefully with small volumes of 0.32 M sucrose, and then suspended in 0.25 M sucrose-0.002 M β-mercaptoethanol solution, to give a dispersion equivalent to 500 mg of the original fresh tissue weight per milliliter. When the base-exchange reaction was studied, brain microsomes were prepared in an EDTA-containing medium, as already reported (12).

The *de novo* synthesis of 1,2-diacyl-*sn*-GPC and 1,2-diacyl-*sn*-GPE was studied from cytidine-5'-diphosphate choline (CDPC) and cytidine-5'-diphosphate ethanolamine (CDPE), respectively, following previous work (5, 6). Incorporation was assayed according to already published procedures (5, 6). The base-exchange reaction was determined as reported elsewhere (12, 15). Protein was determined according to Lowry et al. (16), with crystalline bovine serum albumin as a standard.

III. RESULTS

The effect of EFA deficiency was examined in *de novo* synthesis of phospholipids in brain by incubating microsomes of control and deficient animals *in vitro* with labeled CDPC or CDPE without adding exogenous diacyl glycerols. As known, incubation with CDPC or CDPE overcomes the rate-limiting steps of phospholipid biosynthesis (3, 4, 7), thus producing much higher incorporation rates of labeled precursors into lipid than incubation with phosphorylcholine or phosphorylethanolamine, even in the absence of added diacyl glycerols (4–6). Omitting the exogenous diacyl glycerols in our incubation system confined the possible changes of enzymic activity to variations of endogenous diacyl glycerols resulting from dietary conditions.

Table 1 shows a substantial increase of lipid synthesis in the brain microsomes obtained from the animals fed the fatty acid-deficient diet, as compared to the control values. As a whole, the incorporation rate of CDPC into CPG increases by a value of about 70%. By using thin-layer and column chromatographic techniques (15), all the radioactivity detected in the CPG fraction was found in 1,2-diacyl-*sn*-GPC.

In contrast to the results of Table 1, Table 2 indicates that no significant change of EPG synthesis from labeled CDPE is observed during the deficiency treatment. In no case is the incorporation of labeled CDPE into the phospholipid different from control values. This is confirmed by the isolation of 1,2-diacyl-*sn*-GPE from the radioactive EPG fraction by thin-layer and column chromatography.

The base-exchange reaction mechanism of serine and ethanolamine was

TABLE 1. *Synthesis of choline phosphoglycerides in vitro in brain microsomes of control (C) and fatty acid-deficient (FAD) rats*

Treatment	Activity[a]
C	13.2
	14.1
	13.0
	15.6
	10.6
	14.5
FAD	19.4
	21.7
	20.6
	19.9
	17.6
	22.0

[a] Results expressed as nmoles labeled lipid × mg protein^{-1} × 30 min^{-1}. Each datum is the mean of duplicate estimations.

Incubations carried out for 30 min at 37°C as reported elsewhere (6), with about 1.5 mg microsomal protein. Incorporation rates determined as reported previously (5, 6). See the text for details.

TABLE 2. *Synthesis of ethanolamine phosphoglycerides in vitro in brain microsomes of control (C) and fatty acid-deficient (FAD) rats*

Treatment	Activity[a]
C	10.3
	11.9
	9.4
	11.2
	10.8
	7.9
FAD	10.1
	8.9
	12.0
	7.3
	11.1
	10.4

[a] Results expressed as nmoles labeled lipid $\times$ mg protein^{-1} $\times$ 30 min^{-1}. Each datum is the mean of duplicate estimations.

Incubations carried out for 30 min at 37°C as reported elsewhere (5), with 1.5 mg microsomal protein. Incorporation rates determined as reported previously (5, 6). See the text for details.

TABLE 3. *Exchange of serine and ethanolamine in brain microsomes of control (C) and fatty acid-deficient (FAD) rats*

Treatment	Labeled precursor	Activity[a]
C	L-Serine	4.5 ± 0.9 (6)
	Ethanolamine	6.2 ± 1.1 (6)
FAD	L-Serine	6.0 ± 1.0 (6)
	Ethanolamine	7.9 ± 1.4 (6)

[a] Results are expressed as nmoles labeled lipid $\times$ mg protein^{-1} $\times$ 30 min^{-1} $\pm$ S.E.M. Number of experiments in parentheses. Each experiment represents the mean of duplicate estimations.

Incubations were performed according to Porcellati et al. (12) for 30 min at 37°C, with 1.5 mg microsomal protein, in the presence of 10 mM Ca^{2+} ions. See the text for other details.

also examined. Rat brain microsomes were incubated separately with labeled L-serine or ethanolamine in the presence of suitable amounts of Ca^{2+} ions (12), and the labeled 1,2-diacyl-*sn*-GPS or 1,2-diacyl-*sn*-GPE resulting from the exchange between the incubated free base and the endogenous microsomal phospholipids was isolated (12, 15). Table 3 indicates that brain microsomes of the deficient animals exchange both nitrogenous bases at a rate which is 30% higher than that of controls.

In connection with the data presented in Tables 1 through 3, it must be mentioned that the results with the controls (rats fed a fatty acid-deficient diet supplemented with 5% corn oil) were not different from those of parallel experiments carried out with control groups of rats fed Purina laboratory chow for the same period of time.

IV. DISCUSSION

We have reported in previous work (3, 7, 8) that the fatty acyl composition of membrane phospholipids influences to a different extent in brain particulate fractions the stimulation of the cytidylyltransferase reaction linked to phospholipid biosynthesis. The finding that fatty acid deficiency produced *in vivo* in laboratory rats influences phospholipid synthesis in brain microsomes *in vitro* may be considered, therefore, as a consequence of changes in the molecular structures of membrane phospholipids due to the dietary conditions, deriving from modifications in their fatty acid profiles (10, 11). The failure of EFA deficiency to modify the synthesis of 1,2-diacyl-*sn*-GPE is in agreement with the finding that no stimulation of 1,2-diacyl-*sn*-GPE formation by a variety of added phospholipids occurs *in vitro* (4–6).

It has been previously reported (7) that dienoic diacyl glycerol species in brain are utilized more efficiently than tetraenoic and hexaenoic species for lecithin synthesis *in vitro*. The finding that the synthesis *in vitro* of lecithin by endogenous diacyl glycerols and CDPC is higher in the group of deficient rats than in the controls (Table 1) may be in line with the results of Galli et al. (10), who found a decrease in the tetraenes and hexaenes of brain lipids and a relative proportional increase in the total of the dienes under comparable conditions.

The base-exchange enzymic system for the synthesis of phospholipids was reported to be affected by the acyl composition of the exchanging phospholipids (7, 9). The results of Table 3 are in good agreement with this observation, since a significant, although moderate, change in enzyme activity was observed in EFA deficiency which alters the fatty acid profile of membrane phospholipid components.

V. SUMMARY

An EFA deficiency was induced in female rats up to 6 months of age. Enzymic activities involved in lecithin and phosphatidylethanolamine synthesis and in base-exchange reactions, which occur specifically in membranes, were examined in brain microsomes of both deficient and control groups *in vitro*.

The rate of synthesis of lecithin from cytidine diphosphate choline and endogenous diacyl glycerols was noticeably higher in the deficient rats than in the controls, whereas that of phosphatidyl-ethanolamine was unaffected. Both rates of exchange of serine and ethanolamine were significantly increased by the fatty acid deficiency.

The results are probably due to changes in the fatty acyl profile of some membrane phospholipids induced by the EFA deficiency.

ACKNOWLEDGMENT

The author is grateful to the Consiglio Nazionale delle Ricerche (Rome) for the financial support of this research (contract no. 71.00876.04).

REFERENCES

1. Fiscus, W. G., and Schneider, W. C. (1966): *J. Biol. Chem.*, 241:3324.
2. Chojnacki, T., Radomińska-Pyrek, A., and Korkzybsky, T. (1967): *Acta Biochim. Polon.*, 14:383.
3. Porcellati, G., and Arienti, G. (1970): *Brain Res.*, 19:451.
4. Porcellati, G., and Pirotta, M. G. (1970): *Enzymology*, 38:351.
5. Porcellati, G., Biasion, M. G., and Pirotta, M. G. (1970): *Lipids*, 5:734.
6. Giorgini, D., De Medio, G., and Porcellati, G. (1972): *Il Farmaco*, 27:3.
7. Porcellati, G. (1972): In: *Advances in Enzyme Regulation*, Vol. X, edited by G. Weber. Academic Press, New York.
8. Porcellati, G. (1973): In: *Metabolic Regulation and Functional Activity in the Central Nervous System*, edited by E. Genazzani and H. Herken, Springer-Verlag, Berlin (*in press*).
9. Porcellati, G. (1972): In: *Role of Membranes in Secretory Processes*, edited by L. Bolis, R. D. Keynes, and W. Wilbrandt. North-Holland, Amsterdam.
10. Galli, C., White, H. B., Jr., and Paoletti, R. (1970): *J. Neurochem.*, 17:347.
11. Sun, G. Y. (1972): *J. Lipid Res.*, 13:56.
12. Porcellati, G., Arienti, G., Pirotta, M. G., and Giorgini, D. (1971): *J. Neurochem.*, 18:1395.
13. Porcellati, G., and di Jeso, F. (1971): *Membrane-Bound Enzymes*, edited by G. Porcellati and F. di Jeso. Plenum Press, New York.
14. Thomasson, H. J. (1969): *Nutr. Rev.*, 27:85.
15. Gaiti, A., Goracci, G., De Medio, G., and Porcellati, G. (1972): *FEBS lett.*, 27:116.
16. Lowry, O. H., Rosebrough, N. J., Farr, A. L., and Randall, R. J. (1951): *J. Biol. Chem.*, 193:265.

Dietary Lipids and Postnatal Development
Raven Press, New York © 1973

Lipid Interactions with Brain, Body, and Behavior

D. A. Levitsky

Graduate School of Nutrition and Department of Psychology, Cornell University, Ithaca, New York 14850

Malnutrition experienced during the first 3 weeks of postnatal life produces changes in lipid metabolism in brain and peripheral tissues in the rat. The malnutrition may be caused by administering to the lactating dam a low-protein diet or a complete diet in restricted amounts, or it may be produced by restricting the total amount of the time pups spend feeding from the dam. Some of the effects most commonly observed in the brain are a decrease in total brain cholesterol (1–4) and in myelin content (5–10). Peripherally, rats subjected to such treatment show a decrease in all body components, the largest decrease being in fat content (11). Moreover, the amount of peripheral fat of rats subjected to such treatments, but allowed to recover, continues to show this decrease (11–13).

Malnutrition during this period of life also produces long-term alterations in behavior. These changes are manifested in delayed motor development (14) and decreased exploratory behavior (15). Many behavioral abnormalities continue to be displayed long after nutritional rehabilitation (16–20).

These data raise an interesting question of relationship; that is, are any of these behavioral changes related to changes in brain and/or body lipids? As is readily apparent, the methodology for lipid analysis is far ahead of the methodology for behavioral analysis. Moreover, the kinds of questions asked by biochemists have not stimulated the interest of the behaviorists, at least not until fairly recently. Thus, when one poses the question of a relationship between lipids and behavior to the behaviorist, he cannot fall back upon a well-worked-out procedure as the biochemist would do in isolating a new compound. He must first ask the fundamental question of association, then proceed to reduce this to possible causal mechanisms.

Let us begin to examine this question by reviewing some of the results from experiments concerned with early malnutrition and their long-term behavioral effects in an attempt to specify what kinds of behavior may be affected by alterations in early nutrition. Our procedure for producing malnutrition in all the following studies is as follows. Beginning on day 1 post-partum, the dam is placed on a low-protein diet (12% casein). This procedure has its primary effect in reducing the amount of milk the lactating dam can supply. The pups are weaned to a very low-protein diet (3% casein) for the next 4 weeks of life. The animals are then given the control diet (25% casein) and allowed to recover for at least 10 weeks before any behavioral observation is made.

One of the most obvious characteristics of these animals is that they are smaller in size. Figure 1 shows the growth curve from a typical experiment. With this kind of dietary procedure we typically find about a 20 to 25% reduction in total body weight in animals at approximately 20 weeks of age. This difference continues throughout life. Carcass analysis shows that all body constituents are reduced. However, when the data are expressed as a percentage, the previously malnourished animals show significantly less total body fat and more water than controls (11, 13).

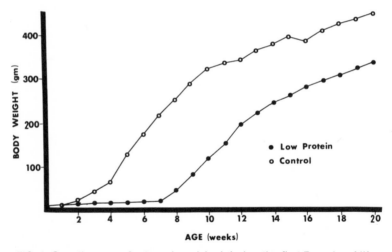

FIG. 1. Growth curve of rats malnourished during the first 7 weeks of life.

Behaviorally, the easiest difference to detect due to early malnutrition is in feeding and drinking behavior (see Table 1). The previously malnourished rats consumed more food and water per unit body weight (or metabolically active tissue) than their well-fed counterparts. So far we have not been able to detect any physiological impairment which may be responsible for these differences, such as decreased intestinal absorption or impaired ADH secretion.

TABLE 1. *Behavioral effects of early malnutrition*

	Control	Malnourished	p values
Food consumed			
g/kg body weight/day	4.2 ± 0.12(18)[a]	5.2 ± 0.14(18)	<0.001
g/kg (body weight) 3/4/day[b]	0.205 ± 0.003(18)	0.224 ± 0.005(18)	<0.01
Percent food spilled in 1-hr test	3.5 ± 0.58(12)	24.6 ± 4.3(12)	<0.001
Water consumed			
ml/kg body weight/day	6.25 ± 0.27(16)	8.60 ± 0.45(15)	<0.001
ml/kg (body weight) 3/4/day[b]	0.285 ± 0.015(16)	0.374 ± 0.025(15)	<0.01
Spontaneous locomotion			
(squares crossed)	51.9 ± 5.4(15)	76.9 ± 7.3(17)	<0.01
Reaction to loud noise (percent			
reduction in movement)	56.0 ± 2.6(7)	74.3 ± 5.4(7)	<0.01
Passive avoidance (reciprocal			
latency)	0.095 ± 0.016(7)	0.020 ± 0.016(7)	<0.01
Sidman avoidance			
(bar press/min)	3.97 ± 0.20(8)	5.59 ± 0.45(8)	<0.01

[a] Mean and standard error, numbers of animals within parenthesis.
[b] Intake expressed as a percentage of the active metabolic mass.

When one examines feeding behavior more closely, another important behavioral characteristic emerges. If the time available for feeding is restricted to 1 hr/day, the previously malnourished rats become very excited at feeding time, and when offered a dry powdered diet they will spill considerably more food than the controls (see Table 1). Moreover, when placed in an operant situation where a response such as bar pressing is required in order to receive a food pellet, the previously malnourished rats are significantly more sensitive to food deprivation conditions than their controls.

The increased behavioral activity can be observed in many other situations (see Table 1). When the previously malnourished animals are placed in a large open area, they exhibit more "emotional"-like responses, such as very fast, flighty movements (locomotion) and more defecation, urination, and vocal distress sounds. After a period of time the animals quiet down (adapt), but this increase in emotional behavior can be redisplayed if the animal is frightened again. We have used a very loud noise to frighten the animal and have measured the amount of disruption in on-going behavior, and have repeatedly observed that the previously malnourished animals are more affected by the frightening stimulus than controls.

Behavioral testing in other situations has confirmed our observations. When the previously malnourished rat is placed on a small platform elevated from grid floor and given an electric shock as soon as the animal reaches the floor, it will remain considerably longer on the platform before stepping down the second time, when compared to controls (passive avoidance). The previously malnourished rat will press a bar more frequently than his well-nourished control in an attempt to delay an electric shock (Sidman avoidance). Once the animals learn a response to avoid an electric shock, it takes them considerably longer to extinguish the response. Thus, emotional reactivity is one dimension of behavior which appears to be increased as a result of malnutrition experienced early in life.

Learning is a very important aspect of behavior in which there is considerable interest. However, to measure learning is not as easy or obvious as one might think. One is not able to measure directly learning *per se*, but only the effects of learning on performance. Unfortunately, there exists a whole host of other factors which also affect performance. Thus, in order to examine the effect of some variable, such as malnutrition, on learning, one must control or equate all other factors affecting performance and only then can one conclude an effect on learning. Most of the studies which have attempted to examine the effect of dietary manipulation on learning have not taken this problem into consideration, and many of their conclusions that malnutrition produces an effect on learning are unjustified (21).

We have attempted to look at learning behavior in the previously malnourished rat in a situation in which we controlled for one of the most potent variables affecting performance: motivation. We first trained rats to press a bar in a Skinner box in order to obtain food. The animals were all fasted, but we varied the degree of fasting by adjusting the daily ration of food so that all animals exhibited the same rate of pressing, indicating approximately the same degree of motivation. Bar-pressing rate is a good index of food motivation (22). It was only after we fixed the level of motivation that we introduced a visual discrimination task. The animals had to learn to press another set of bars which corresponded to a pattern of illuminated lights. We found no difference in the rate of learning this discrimination problem as a function of early malnutrition. We then looked at many variations on this problem, but could find no difference in the rate of learning of any task in which motivation was held constant.

Summarizing, what we observed is that malnutrition suffered early in life produces long-term, persistent behavioral abnormalities. The behavioral abnormality expresses itself primarily as an increase in behavioral responsiveness to environmental manipulations, which is most evident under conditions of stress. Learning ability, however, does not seem to be affected by this early dietary treatment, although cumulative cognitive deficits can be shown to occur under particular conditions (23).

Thus, a severe dietary restriction which occurred early in life produces an animal which is more reactive and contains less lipid. When one examines the literature, it becomes apparent that a number of other experimental manipulations also yield an inverse relationship between lipid content (composition) and behavioral activity. This is shown in Table 2. There are three important points to be made from this table. First, in every situation listed there exists a perfect inverse relationship between lipid content and behavioral activity. Second, the manipulation of behavioral activity *produces* a corresponding change in lipid content, either directly by physically restricting movement (26) or by forced exercise (27), or indirectly via the use of central nervous system stimulants (35) or depressants (37).

TABLE 2. *Relationship between fat content and behavior*

Procedure	Fat	Behavior
1. Early malnutrition	− (11)	+ (19)
2. Genetics	+ (24)	− (25)
3. Environmental restriction	+ (26)	−
4. Exercise	− (27)	+
5. Cold	− (28)	+ (29)
6. Ventro medial lesion of hypothalamus	+ (30)	− (31)
7. Fasting	− (32)	+ (33)
8. Force feeding	+ (34)	− (34)
9. Stimulants	− (35)	+ (36)
10. Depressants	+ (37)	− (38)
11. Estrogen	− (39)	+ (39)
12. Progesterone	+ (39)	− (39)
13. Age	+ (40)	− (41)

Finally, procedures which directly increase fat stores, such as forced feeding (34) or hypothalamic lesions (30), *produce* the same corresponding changes in behavior. The results of the Cohn and Joseph studies (34) are particularly important in this respect. They point out that even after the forced-feeding procedure was terminated, behavioral activity remained low until body weights (fat content) decreased to approximately control levels, after which activity increased to control values. It also appears that the activity level of hypothalamically lesioned animals is depressed coincidentally with increasing fat content (31). Hoebel and Thompson (42) report self-stimulation rate of rats to be inversely proportional to body weight increases due to forced feeding.

It is not clear, however, whether intake of dietary fats produces predictable changes in behavioral activity. Frankova (43) reports several experiments in rats where increases in the amount of fat in the diet actually increases various parameters of behavioral activity. However, she also reports that these diets did not produce changes in body weight. Others, however, have reported a decrease in behavioral activity with increases in dietary fat (44). Alexander and Kopeloff (45) report data showing that feeding of high-cholesterol diets depresses the susceptibility of rats to chemically induced seizures. Others have suggested a high-fat and high-cholesterol diet for the treatment of epilepsy (46, 47).

Our current state of knowledge does not allow a sufficient explanation of these phenomena. However, the consistency of the relationship found in the study of these many, seemingly unrelated, variables strongly suggests that lipids may play a far greater role in the modulation of behavior than has heretofore been suspected.

REFERENCES

1. Culley, W. J., and Mertz, E. T. Z. (1965): *Proc. Soc. Exp. Biol. Med.*, 118:233.
2. Guthrie, H. A., and Brown, M. L. (1968): *J. Nutr.*, 94:419.
3. Culley, W. J., and Lineberger, R. O. (1968): *J. Nutr.*, 96:375.
4. Howard, E., and Granoff, D. M. (1968): *J. Nutr.*, 95:111.
5. Dobbing, J. (1964): *Proc. Roy. Soc. (Biol.)*, 159:503.
6. Benton, J. W., Moser, H. W., Dodge, P. R., and Larr, S. (1966): *Pediatrics*, 38:801.
7. Davison, A. M., and Dobbing, J. (1966): *Brit. Med. Bull.*, 22:40.
8. Chase, H. P., Dorsey, J., and McKhann, G. M. (1967): *Pediatrics*, 40:551.
9. Dobbing, J., and Widdowson, M. (1965): *Brain*, 88:357.
19. Dobbing, J. (1968): In: *Applied Neurochemistry*, edited by A. N. Davison and J. Dobbing. Blackwell Press, Oxford.
11. Heggeness, F. W. (1961): *Am. J. Physiol.*, 201:1044.
12. Knittle, J. L., and Husch, J. (1967): *Clin. Res.*, 15:323.
13. Barnes, R. H., and Kwong, E. (*in press*): *J. Nutr.*
14. Smart, J. L., and Dobbings, J. (1971): *Brain Res.*, 28:85.
15. Frankova, S., and Barnes, R. H. (1968): *J. Nutr.*, 96:477.
16. Barnes, R. H., Cannold, S. R., Zimmerman, R. R., Simmons, H., MacLeod, R. B., and Krook, L. (1966): *J. Nutr.*, 89:399.
17. Guthrie, H. A. (1968): *Physiol. Behav.*, 3:619.
18. Frankova, S., and Barnes, R. H. (1968): *J. Nutr.*, 96:485.
19. Levitsky, D. A., and Barnes, R. H. (1970): *Nature*, 225:468.
20. Levitsky, D. A., and Barnes, R. H. (1972): *Science*, 176:68.
21. Levitsky, D. A., and Barnes, R. H. (*in press*): Proceedings from Conference on the Assessment of Tests of Behavior from Studies of Nutrition in the Western Hemisphere.
22. Collier, G., Levitsky, D., and Squible, R. L. (1967): *J. Comp. Physiol. Psychol.*, 64:68.
23. Levitsky, D. A. (*in press*): *Proceedings of the IV International Congress on Nutrition.*
24. Zacker, T. F., and Zacker, L. M. (1963): *J. Nutr.*, 80:6.
25. Mayer, J. (1953): *Science*, 117:504.
26. Engle, P. J. (1949): *Proc. Soc. Exp. Biol. Med.*, 72:604.
27. Parizkova, J., and Stankova, L. (1964): *Brit. J. Nutr.*, 18:325.
28. Schmidt, P., and Widdowson, E. M. (1967): *Brit. J. Nutr.*, 21:457.
29. Stevenson, J. A. F. (1951): In: *Cold Injury*, Third Conference, Josiah Macy, Jr., Foundation, New York.
30. Montamurro, P. G., and Stevenson, J. A. F. (1960): *Am. J. Physiol.*, 198:757.
31. Hetherington, A. W., and Ranson, S. W. (1942): *Am. J. Physiol.*, 136:609.
32. Lee, M., and Lucia, S. P. (1961): *J. Nutr.*, 74:243.
33. Campbell, B. A., and Sheffield, F. D. (1953): *J. Comp. Physiol. Psych.*, 46:320.

34. Cohn, C., and Joseph, D. (1962): *Yale J. Biol. Med.*, 34:598.
35. Stowe, F. R., Jr., and Miller, A. R., Jr. (1957): *Experientia*, 13:144.
36. Stein, L. (1964): In: *Ciba Foundation Symposium*, p. 91.
37. Stolerman, I. P. (1967): *Nature*, 215:1518.
38. Cook, L., and Kelleher, R. T. (1963): *Ann. Rev. Pharmac.*, 3:205.
39. Wade, G. N. (1972): *Physiol. and Behav.*, 8:523.
40. Spray, C. M., and Widdowson, E. M. (1950): *Brit. J. Nutr.*, 4:332.
41. Furchtgott, E., Wechkin, S., and Dees, J. (1961): *J. Comp. Physiol. Psychol.*, 54:386.
42. Hochel, B. G., and Thompson, R. D. (1969): *J. Comp. Physiol. Psychol.*, 68:536.
43. Frankova, S. (1963): *Act. Ner. Sup.*, 3–4:71.
44. Smith, E. A., and Conger, R. M. (1944): *Am. J. Physiol.*, 142:663.
45. Alexander, G. T., and Kopeloff, L. M. (1971): *Exp. Neurol.*, 32:134.
46. Aird, R. B., and Gurchot, C. (1939): *Arch. Neurol. Psychiat.*, 42:491.
47. Dekaban, A. S. (1966): *Arch. Neurol.*, 15:177.

Dietary Lipids and Postnatal Development
Raven Press, New York © 1973

The Mechanism of Drug Secretion into Milk

Folke Rasmussen

The Institute of Pharmacology and Toxicology, Royal Veterinary and Agricultural University, Copenhagen, Denmark

The excretion of drugs into milk by humans and animals, especially ruminants, is of practical as well as of academic interest. For the suckling, breast milk is the only or the main part of the diet, and throughout life, in most parts of the world, cow's or goat's milk are chief components of the human diet either raw or as processed products. Hidden contact with drugs through these important food items has caused harmful reactions in consumers.

The excretion of drugs into milk from humans and animals has been reviewed by Reed (1), Déchavanne (2), Kolda (3), Joachimovits (4), Dreyfus-Sée (5), Sapeika (6, 7), Burn (8), Albright et al. (9), Marth (10), Reusse (11), Knowles (12), Rasmussen (13, 14), Hüter and Zehentbauer (15), Dolby and Kronberg (16), and Catz and Giacoia (17). The mechanism of the mammary excretion of drugs has been studied for the last 15 years and is now quite well elucidated (14).

I. THE MAMMARY GLAND

The mammary gland consists of irregular lobes provided with excretory ducts radiating from the mammary papilla or from the teat and the gland cistern in ruminants. The wall of the secretory portions, the alveolar ducts, and the alveoli consist of a basement membrane, a layer of myoepithelial cells and, on the internal surface of the resting gland, a row of low columnar glandular cells. During lactation the shape of the glandular cells fluctuates from cylindrical or conical to flat. This biological membrane separates the extracellular fluid from the secretion. Water and various compounds dissolved in the extracellular fluid can pass through the membrane, in which synthesis also takes place. The milk therefore contains some components which are found preformed in the extracellular fluid, and others which are synthesized during the process of secretion in the glandular epithelium.

231

Drugs belong to the former group, although it has recently been shown that some drugs are metabolized in the mammary tissue, for example, sulfanilamide, which is N^4-acetylated (18).

II. PASSAGE OF DRUGS FROM BLOOD PLASMA TO MILK

A few minutes after the rapid intravenous injection into cows of an aqueous solution of a sulfonamide, the drug is demonstrable in milk samples (13). The sulfonamide levels in simultaneously drawn samples of blood and milk from cows at intervals of hours after intravenous injection of therapeutic doses of different sulfonamides revealed differences with regard to the rate of mammary excretion. However, these types of analyses, done during falling blood concentrations, give no absolutely reliable expression of the relationship between the level of drugs in milk and plasma. This requires a determination of the distribution at equilibrium, obtainable by continuous intravenous infusion (19). In such experiments it was observed that the concentrations of antipyrine in milk and blood plasma were almost identical, although the concentration of sulfadimidine in milk was one-third the concentration in blood plasma. The concentration of trimethoprim in milk, on the other hand, was a little higher than in blood plasma (Fig. 1).

The protein-bound fraction of drugs cannot penetrate biological membranes, however, so the ratio of the concentration in ultrafiltrate of milk to that in ultrafiltrate of blood plasma (ratio M.Ultr./P.Ultr.) was calculated and found to be 0.6, 1.0, and 2.5 for sulfadimidine, antipyrine, and trimethoprim, respectively (Fig. 2). To decide whether the blood plasma level influenced the ratio M.Ultr./P.Ultr., the ratio was estimated at different concentrations in blood plasma; from Fig. 3 it can be seen that the ratio is independent of the blood plasma level within the range of concentrations studied (19–21). Similar results were published by Sisodia and Stowe (22). These findings are consistent with the view that excretion into milk takes place by diffusion.

A. Dependence on Drug Ionization

The view that the degree of ionization influences the diffusion equilibrium across a biological membrane presupposes the presence of a pH gradient across the membrane. Since the pH of the blood plasma is between 7.4 and 7.6 in goats and cows, and the pH of milk is between 6.6 and 6.8 (7.0 in human milk), the degree of ionization of weak acids and bases will differ in the two fluids. If we accept that it is the un-ionized fraction which diffuses through membranes (23, 24), a basis for calculation of the equilibrium can

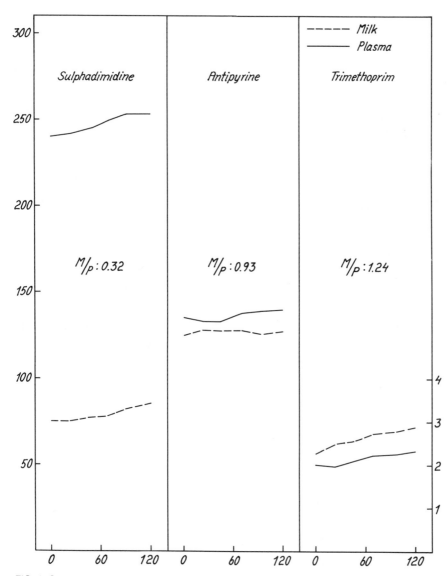

FIG. 1. Concentrations of drugs in milk and plasma during continuous infusion. Ordinate: concentration (μg/ml); abscissa:time (min).

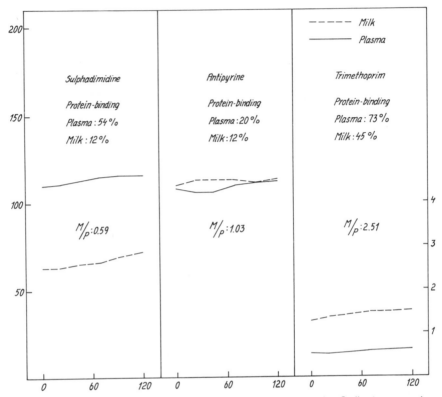

FIG. 2. Concentrations of drugs in ultrafiltrates of milk and plasma. Ordinate:concentration (μg/ml); abscissa:time (min).

be established. The hypothesis that diffusion depends on the un-ionized form of the drug was strengthened by the results of a series of experiments during which large pH variations were found in the milk samples, probably owing to frequent strippings (19).

Such an experiment is shown in Fig. 4. Continuous infusion of the acid sulfadoxine and the base trimethoprim gave constant blood plasma levels of 14 to 17 μg/ml and 0.8 to 0.9 μg/ml, respectively. The concentrations of drug in simultaneously drawn milk samples varied appreciably, but the concentrations of sulfadoxine were in all cases lower than in blood plasma, although the concentrations of trimethoprim in milk were in all cases higher than in blood plasma. However, the figure shows that there was a direct relationship between the sulfonamide concentration and the pH of milk

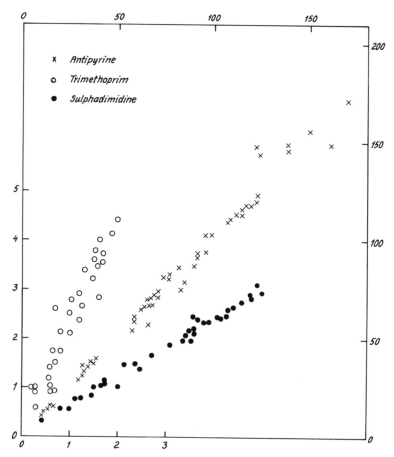

FIG. 3. Relationship between concentrations of sulfadimidine, antipyrine, and trimethoprim in ultrafiltrates of milk and plasma. Ordinate on the left: concentration of trimethoprim in ultrafiltrate of milk (μg/ml); ordinate on the right: concentration of sulfadimidine
and antipyrine in ultrafiltrate of milk (μg/ml). Abscissa above: concentration of sulfadimidine and antipyrine in ultrafiltrate of plasma (μg/ml); abscissa below: concentration
of trimethoprim in ultrafiltrate of plasma (μg/ml).

samples, whereas the concentration of trimethoprim decreases when pH
increases. When changes of pH in milk influence the concentrations of sulfonamide and trimethoprim in opposite directions while the concentration of
the fully un-ionized urea is constant, this indicates that the degree of ionization influences the distribution of sulfadoxine as well as trimethoprim between ultrafiltrates of milk and blood plasma (21). A similar relation between

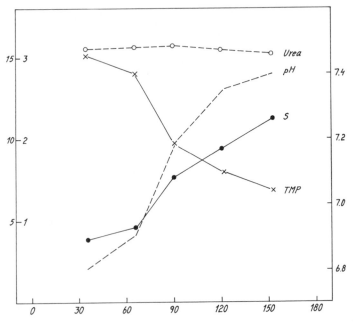

FIG. 4. Concentrations of urea, sulfadoxine, and trimethoprim in ultrafiltrates of milk at different pH values in milk. Ordinate on the left: concentration of sulfadoxine (1 to 15 μg/ml) and trimethoprim (1 to 3 μg/ml); ordinate on the right: the pH of the milk; abscissa: time (min) o ---- o, Urea; x ——— x, Trimethoprim; • ——— •, Sulfadoxine; --------, pH of the milk.

the mammary excretion of weak electrolytes and spontaneous pH variations in milk was also seen in the case of other sulfonamides (19, 13), the weak acid penicillin, and the weak base erythromycin (25).

After intramammary infusion of bicarbonate buffers into one gland of a cow udder, Miller et al. (26) compared the experimental and theoretical ratios M.Ultr./P.Ultr. in normal as well as "alkaline glands" and found good agreement between the experimental and the theoretical ratios at pH variations in milk from 6.6 to 8.1.

The fact that the concentration of sulfonamide in milk is lower than in blood plasma whereas the concentration of trimethoprim in milk is higher than in blood plasma, together with the fact that the concentrations of the two drugs change in opposite directions when pH in milk increases, indicates that only the un-ionized fraction diffuses through the biological membrane. The amount of un-ionized drug in ultrafiltrates of milk and blood plasma can be calculated by means of the Henderson-Hasselbalch equation:

$$\log \frac{U}{I} = pK_a - pH \qquad \text{for an acid}$$

and

$$\log \frac{I}{U} = pK_a - pH \qquad \text{for a base}$$

When pK_a and pH values are known, the ratio between concentrations of un-ionized (U) and ionized (I) drug, and thus the percentage of un-ionized drug, can be calculated.

The distribution of sulfadimidine, antipyrine, and trimethoprim between ultrafiltrates of milk and blood plasma has been shown previously (Fig. 2). Whereas the base antipyrine (pK_a 1.4) is un-ionized in both milk and blood plasma, the acid sulfadimidine (pK_a 7.4) and the base trimethoprim (pK_a 7.6) will be present in partly ionized form in the two fluids. The calculated

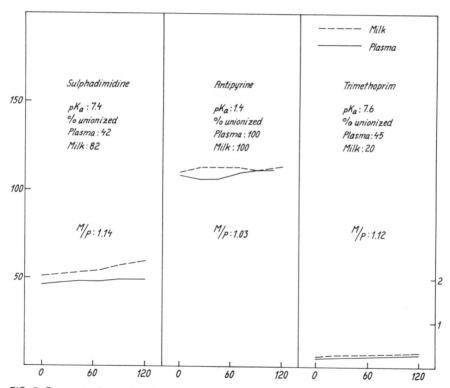

FIG. 5. Concentrations of un-ionized drugs in ultrafiltrates of milk and plasma. Ordinate: concentration (μg/ml); abscissa: time (min).

concentrations of un-ionized drugs in ultrafiltrates of milk and blood plasma are seen in Fig. 5. Here the almost identical curves show that equilibrium is obtained via the mammary gland epithelium between the un-ionized drug in milk and blood plasma. This is in agreement with the theory that the passage through the mammary gland epithelium occurs by diffusion of the non-protein-bound and the un-ionized fractions.

B. Experimental and Theoretical Ratios M.Ultr./P.Ultr.

On the assumption that at equilibrium the un-ionized fraction of a partially ionized drug is the same in ultrafiltrate of milk as in ultrafiltrate of blood plasma, the ratio between the total concentrations of the drug in ultrafiltrates of milk and blood plasma can be calculated.

TABLE 1. *Excretion of drugs by milk in experiments on cows and goats with a constant level of drug in blood plasma*

Drug	pK_a	Theoretical M.Ultr./P.Ultr.	Experimental M.Ultr./P.Ultr.	References
Acids				
Penicillin	2.7	0.16	0.25	(25)
Sulfadimethoxine	6.0	0.19	0.20	(22)
Sulfamethoxazole	6.0	0.22	0.18	(27)
Sulfadoxine	6.1	0.21	0.21	(27)
	6.1	0.4	0.5	(21)
Phenobarbital	7.2	0.5	0.7	(13)
	7.2	0.5	0.85	(28)
Sulfadimidine	7.4	0.51	0.59	(19)
(sulfamethazine)	7.37	0.58	0.51	(22)
	7.4	0.57	0.55	(29)
	7.4	0.67	0.68	(27)
Barbital	7.8	0.8	0.9	(13)
Pentobarbital	8.0	0.9	1.1	(13)
	8.0	0.85	0.91	(28)
Sulfanilamide	10.4	1.00	0.97	(19)
	10.4	1.00	1.00	(22)
	10.4	1.00	1.05	(26)
	10.4	1.00	1.06	(27)
Bases				
Antipyrine	1.4	1.00	1.00	(20)
	1.4	1.00	0.95	(26)
	1.4	1.00	1.01	(30)
Lincomycin	7.6	2.8	3.9	(31)
Trimethoprim	7.6	3.9	3.7	(32)
	7.6	1.9	2.3	(21)
Quinine	8.4	4.0	4.2	(26)
Ephedrine	9.6	7.9	7.9	(26)

Conversion of the Henderson-Hasselbalch equation gives, for an acid,

$$R = \frac{M.Ultr.}{P.Ultr.} = \frac{1 + 10^{(pH_m - pK_a)}}{1 + 10^{(pH_b - pK_a)}}$$

and for a base,

$$R = \frac{M.Ultr.}{P.Ultr.} = \frac{1 + 10^{(pK_a - pH_m)}}{1 + 10^{(pK_a - pH_b)}}$$

By means of the measured pH of milk (pH_m) and blood (pH_b) and the pK_a value of the drug, M.Ultr./P.Ultr. ratios have been calculated. Some examples are given in Table 1, where the theoretically calculated ratios are compared with those found experimentally under equilibrium conditions. For all acids the ratio M.Ultr./P.Ultr. was found to be 1 or lower, whereas for all bases it was found to be 1 or higher. Furthermore, a direct relationship was found between the ratios and the pK_a values, the ratio being highest for drugs with high pK_a values and lowest for drugs with low pK_a values. The table shows a very close agreement between the M.Ultr./P.Ultr. ratios calculated theoretically and those determined experimentally. Further examples are listed by Rasmussen (14).

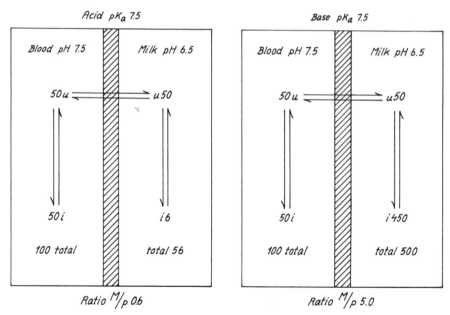

FIG. 6. Schematic representation of the distribution of a weak acid and a weak base across the mammary gland epithelium.

The distribution of a weak acid and a weak base across a lipid membrane was discussed by Brodie and Hogben (33), Schanker (23), and Brodie (34). Rasmussen (13) has applied this concept to the mammary gland epithelium. Figure 6 shows the theoretical distribution of a weak acid on the left and the theoretical distribution of a weak base on the right (13).

III. DISTRIBUTION OF DRUGS IN THE WATER, FAT, AND PROTEIN PHASES OF MILK

Milk is a suspension of fat droplets in an aqueous phase, in which proteins, lactose, and inorganic salts are dissolved; milk is thus a biological fluid in which drugs may be distributed in different phases. Both the aqueous and the lipid phase may act as solvents, and adsorption on proteins is also possible. There are some quantitative differences between breast milk and cow's milk (Table 2) (35), and the differences can account for small quantitative differences in the amounts of drug in breast milk and cow's milk.

TABLE 2. *Composition of mature milk*

Constituent per 100 ml whole milk	Man	Cow
Water, g	88	87
Energy, Cal	65	65
Protein, g	1.2	3.3
Casein, g	0.4	2.8
Lactalbumin, g	0.3	0.4
Lactglobulin, g	0.2	0.2
Carbohydrate, g	7.0	4.8
Fat, g	3.8	3.7
pH	7.0	6.6

In studying the mechanism of the mammary excretion of drugs, their distribution in the various milk phases has shown that the concentration in the aqueous phase of milk, which constitutes the direct contact with the extracellular fluid, need not be identical with the total concentration in the milk. Correction must therefore be made for the fractions of a drug which might be dissolved in milk fat or bound to proteins before the concentration in milk can be compared with that in ultrafiltrate of blood plasma.

A series of experiments with sulfonamides and barbiturates will be described for the purpose of examining the distribution of drugs in the different phases of milk.

A. Distribution in Milk Fat and Skim Milk

After the addition of sulfonamides or barbiturates to skim milk, whole milk, and cream, centrifugation and analysis of the skim milk fractions, which contain less than 0.5% fat, can give information about the distribution in milk fat and in the low-fat skim milk, respectively. The results of such an experiment are given in Table 3, which shows that the concentrations of sulfathiazole, sulfadimidine, and barbital in the skim milk fractions rise with increasing percentages of fat in the starting material, whereas phenobarbital, pentobarbital, and 5-(bicyclo-3,2,1-oct-2-en-2-yl)-5-ethylbarbituric acid are recovered in smaller quantities in the skim milk phase of the high-fat media. This means that lipid soluble drugs are concentrated several times in the milk fat (Table 4).

TABLE 3. *Concentrations of sulfonamides and barbiturates in the skim milk fractions of milk and cream containing 3.6 and 36% fat, respectively*

			Concentration in skim milk			
	Coefficient of lipid solubility	Amounts added (μg/ml)	of whole milk 3.6% (μg/ml)	% of added	of cream 36% (μg/ml)	% of added
Sulfathiazole	0.15	9.7	9.9	102	14.7	152
		47	48	102	69	147
		141	147	104	188	133
Sulfadimidine (sulfamethazine)	3.2	8.6	8.8	102	12.5	145
		44	43	98	60	136
		130	129	99	172	132
Barbital	0.7	45	47	104	66	147
		92	98	107	126	137
		135	135	100	186	138
Phenobarbital	4.8	39	37	95	34	87
		84	81	96	65	77
		122	120	98	101	83
Pentobarbital	28	45	33	73	15	33
		102	72	71	33	32
		147	108	74	60	41
5-(bicyclo-3,2,1-oct-2-en-2-yl)-5-ethylbarbituric acid	>100	47	27	58	20	43
		90	69	77	36	40
		144	113	79	52	36

TABLE 4. *Concentrations of barbiturates in fat-free milk and milk fat after addition of 100 µg barbiturate per ml milk containing 3.6% fat*

Barbiturate	Lipid solubility	Fat-free milk (µg/ml)	Milk fat (µg)	Milk fat (µg/ml)
Barbital	0.7	104	0	0
Phenobarbital	4.8	96	4	111
Pentobarbital	28	73	27	750

This distribution of the barbiturates between the lipid and the aqueous phase is consistent with the lipid solubility expressed by the chloroform/water molecular distribution. For the sulfonamides no such correlation was found between the chloroform/water molecular distribution and the distribution between milk fat and skim milk. The chloroform/water distribution is 0.15 for sulfathiazole and 3.2 for sulfadimidine, but both derivatives seem to occur exclusively in the skim milk phase (13).

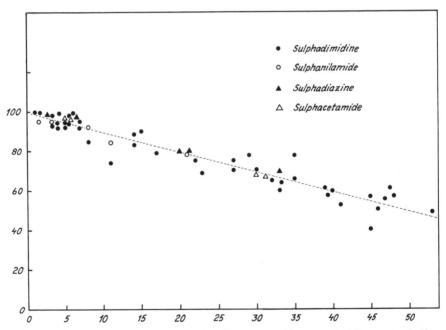

FIG. 7. Concentrations of sulfonamides in milk expressed in percent of the concentration in the skim milk fraction and related to the percentage of fat in the milk. Ordinate: conc. milk/conc. skim milk × 100; abscissa: Percentage of fat in the milk.

This distribution was further borne out by examinations of milk samples from cows and goats given sulfanilamide, sulfadimidine, sulfadiazine, and sulfacetamide, respectively. The milk samples and the high-fat and low-fat fractions prepared from these by centrifugation were analyzed for contents of sulfonamide, and for milk fat by the Röse-Gottlieb method. The concentration of sulfonamide was lower in the high-fat fractions than in the corresponding low-fat fractions. By calculating the determined concentrations as a percentage of the sulfonamide level in the fat-free phases of the milk samples and relating this to the fat contents of the milk samples, we get the diagram shown in Fig. 7. The dotted line indicates the theoretical relationship between the contents of drugs and milk fat when the drugs are insoluble in fat. The diagram shows that the sulfonamide content of the milk samples is reduced at the same rate as the fat content increases. The tested sulfonamides were therefore found in the fat-free phase of the milk (13), and these findings are in agreement with earlier observations (36–39).

B. Milk Protein Binding

The fact that sulfonamides are present in lower concentrations in the high-fat than in the low-fat fraction of milk does not mean that they are only found freely dissolved in the aqueous phase of the milk. By ultrafiltration through cellophane membranes of sulfonamide-containing milk samples, using the technique described by Poulsen (40), protein binding can be demonstrated, which for a number of sulfonamides can be up to 40% of the total sulfonamide present (13, 19).

In general, the degree of binding of drugs to proteins is lower in milk than in plasma from cows; some examples are listed in Table 5. The quanti-

TABLE 5. *Binding of sulfonamides to proteins in milk and blood plasma from cows estimated by ultrafiltration through a cellophane membrane*

	Percent bound in		
Drug	Blood plasma	Milk	References
Sulfacetamide	23	0	(13, 19)
Sulfadiazine	11–14	0–6	(13)
Sulfadimidine (sulfamethazine)	43–73	0–12	(13, 19)
Sulfaethoxypyridazine	90	20–40	(13)
Sulfanilamide	13–22	3–19	(13, 19)
Sulfathiazole	49–69	25	(13, 19)

tative differences in milk proteins between humans and cows (Table 2) can possibly influence the degree of protein binding.

As a result of these considerations, the concentration of a drug in whole milk need not be identical to its concentration in the aqueous phase in milk. Cellophane membranes with a pore size of 20 to 80 Å hold back proteins as well as fat droplets, for example, drugs bound to proteins or dissolved in fat droplets. Ultrafiltration of milk through such a membrane therefore gives an aqueous phase of milk with drug concentrations which are comparable to those in the ultrafiltrate of blood plasma.

IV. CONCLUSION

The mammary gland epithelium, like other biological membranes, acts as a lipid barrier with water-filled pores. These pores seem to be penetrated only by water-soluble, low-molecular-weight compounds such as urea, whereas the water-soluble ionized fractions of drugs with molecular weights above 200 do not pass through the pores.

The un-ionized fraction of the drugs, on the other hand, can penetrate the lipid barrier so that the concentration of the un-ionized drug becomes identical on both sides of the membrane. Since, however, the pH values of the fluids on either side of the mammary gland epithelium are different (7.4 in blood, 7.0 in human milk, and 6.5 to 6.8 in milk from goats and cows), the extent of ionization of most drugs will be different in the two fluids. The nondiffusible, ionized fraction of drugs which are weak acids constitutes a smaller proportion of the total drug content in the ultrafiltrate of milk than in ultrafiltrate of blood plasma. Thus, the sum of the un-ionized and ionized fractions of such drugs will be lower in the ultrafiltrate of milk than in that of blood plasma. On the other hand, the ionized fraction of a drug that is a weak base constitutes a higher percentage of the total content in ultrafiltrate of milk than in ultrafiltrate of blood plasma. The total concentration of such a drug will therefore be higher in the milk than in the plasma.

It is shown that drugs with a high lipid solubility are concentrated in the milk fat, and further, that drugs are to some extent bound to milk proteins.

REFERENCES

1. Reed, C. B. (1908): *Surg. Gynec. Obstet.*, 6:514.
2. Déchavanne, H. (1909): *Du passage des substances médicamenteuses et toxiques dans le lait.* Dissertation, A. Rey, Lyon.
3. Kolda, J., (1926): *Lait*, 6:12, 88, 108, 269.
4. Joachimovits, R. (1929): *Mschr. Geburtsh. Gynäk.* 83:42.
5. Dreyfus-Sée, G. (1934): *Rev. Méd.*, 51:198.

6. Sapeika, N. (1947): *J. Obstet. Gynaec. Brit. Emp.*, 54:426.
7. Sapeika, N. (1959): *S. Afr. Med. J.*, 33:818.
8. Burn, J. H. (1947/48): *Brit. Med. Bull.*, 5:190.
9. Albright, J. L., Tuckey, S. L., and Woods, G. T. (1961): *J. Dairy Sci.*, 44:779.
10. Marth, E. H. (1961): *J. Milk Food Technol.*, 24:36, 70.
11. Reusse, U. (1961): *Dtsch. Tierärztl. Wschr.*, 68:183, 240.
12. Knowles, J. A. (1965): *J. Pediat.*, 66:1068.
13. Rasmussen, F. (1966): *Studies on the Mammary Excretion and Absorption of Drugs.* C. Fr. Mortensen, Copenhagen.
14. Rasmussen, F. (1971): In: *Handbook of Experimental Pharmacology,* edited by O. Eichler, A. Farah, H. Herken, and A. D. Welch, Vol. XXVIII, 1, edited by B. B. Brodie and J. Gillette. Springer-Verlag, New York.
15. Hüter, J., and Zehentbauer, B. (1970): *Übergang von Medikamenten in die Muttermilch und Nebenwirkungen beim gestillten Kind.* Georg Thieme Verlag, Stuttgart.
16. Dolby, J., and Kronberg, L. (1972): *Svensk Farmaceutisk Tidskrift,* 76:197.
17. Catz, C. S. and Giacoia, G. P. (1973): In: *This Volume.*
18. Rasmussen, F., and Linzell, J. L. (1967): *Biochem. Pharmacol.*, 16:918.
19. Rasmussen, F. (1958): *Acta Pharmacol. (Kbh.)*, 15:139.
20. Rasmussen, F. (1961): *Acta Vet. Scand.*, 2:151.
21. Davitiyananda, D., and Rasmussen, F. (*in press*): *Nord. Vet.-Med.*, 25.
22. Sisodia, C. S., and Stowe, C. M. (1964): *Ann. N.Y. Acad. Sci.*, 111:650.
23. Schanker, L. S. (1962): *Pharmacol. Rev.*, 14:501.
24. Schanker, L. S. (1971): In: *Handbook of Experimental Pharmacology,* edited by O. Eichler, A. Farah, H. Herken, and A. D. Welch, Vol. XXVIII, 1, edited by B. B. Brodie and J. Gillette. Springer-Verlag, New York.
25. Rasmussen, F. (1959): *Acta Pharmacol. (Kbh.)*, 16:194.
26. Miller, G. E., Banerjee, N. C., and Stowe, C. M., Jr. (1967): *J. Dairy Sci.*, 50:1395.
27. Jørgensen, S. T. (1972): *Nord. Vet.-Med.*, 24:1.
28. Miller, G. E., Peters, R. D., Engebretsen, R. V., and Stowe, C. M. (1967): *J. Dairy Sci.*, 50:769.
29. Miller, G. E., and Stowe, C. M., Jr. (1967): *J. Dairy Sci.*, 50:840.
30. Miller, G. E., Banerjee, N. C., and Stowe, C. M., Jr. (1967): *J. Pharmacol. Exp. Ther.*, 157:245.
31. Rasmussen, F. (1966): *Acta Vet. Scand.*, 7:97.
32. Rasmussen, F. (1970): *Vet. Rec.*, 87:14.
33. Brodie, B. B., and Hogben, C. A. M. (1957): *J. Pharm. Pharmacol.*, 9:345.
34. Brodie, B. B. (1964): In: *Absorption and Distribution of Drugs,* edited by T. B. Binns. Livingstone, Edinburgh-London.
35. Spector, W. S. (1956): In: *Handbook of Biological Data.* W. B. Saunders, Philadelphia and London.
36. Pinto, L. S. (1938): *J. Am. Med. Ass.*, 111:1914.
37. Rieben, G., and Druey, J. (1942): *Schweiz. Med. Wschr.*, 72:1376.
38. Friesen, B. von (1951): *Acta Obstet. Gynec. Scand.*, 31:suppl. 3:78, 95.
39. Gamklou, R., Rybo, G., and Spetz, S. (1962): *Nord. Med.*, 68:1127.
40. Poulsen, E. (1956): *Renale Clearance-Undersøgelser hos Køer. Renal Clearance in Cows.* C. Fr. Mortensen, Copenhagen.

Dietary Lipids and Postnatal Development
Raven Press, New York © 1973

Drugs and Metabolites in Human Milk

Charlotte S. Catz and George P. Giacoia

Department of Pediatrics, School of Medicine,
State University of New York at Buffalo, Buffalo, New York 14214

I. INTRODUCTION

Several excellent studies have examined in depth the composition of human milk. These have taken into account biochemical and nutritional aspects and have been supplemented with a review of economical and sociological factors related to breast feeding. One of the many well-documented problems of the eighteenth century was the abandonment of infants and the founding of hospices for their care. The insufficient number of available wet nurses contributed significantly to the high mortality rate for that age group, since no alternative means of feeding existed. Great concern was expressed over the feasibility of achieving by other means the same ideal infant nutrition as that attained through human milk. Increasing knowledge and technical advances in the field of nutrition have permitted the human mother to have a choice between human or cow's milk preparations for her infant. Over the last 30 years, surveys in the United States have shown a steady decline in the percentage of babies totally breast fed from 38% in 1946 to 21% a decade later and 18% in 1966. Simultaneously, it was noted that a marked decrease occurred in the number of mothers who continued nursing their infants beyond 2 months of age.

Nevertheless, even if the popularity of bottle feeding has reached a high peak, it has not meant the disappearance of breast feeding. On the contrary, a resurgence is noted presently among the younger generation, and the practice of this ancient art has remained stable in certain population groups.

Attention has been drawn to the possibility that breast milk may contain noxious agents. These were originally attributed to substances present in the diet of the lactating woman, and long, elaborate lists of permitted and prohibited foodstuffs were published. These were mentioned as part of popular beliefs and folklore, and traditionally were handed down from mothers to daughters. Today, the pediatrician is well aware that medications administered to the mother may produce effects in her nursing infant. Many case reports in the literature attest to this fact. The concentration of some

247

of these compounds in maternal plasma and milk and the determination of a milk/plasma ratio have been compiled and published (1–3). These isolated measurements were done at different times after drug intake and, as a consequence, the results are hard to compare. Also, very few studies have addressed themselves to determining the amount of the medications excreted in milk which are actually absorbed by the nursling and to ascertain if the side effects observed can be correlated to the quantity of drug present in the infant. This consideration should not be discussed lightly, since even minimal amounts of drugs which do not produce obvious adverse reactions may cause certain changes in a developing organism.

Possible mechanisms by which these alterations could be produced will be examined.

II. PHARMACOLOGICAL MECHANISMS

A. Cumulative Effect

Drugs bound to fat or tissue proteins are not eliminated rapidly, so a repeat intake of the compound can lead to an accumulation. For instance, free chlorpromazine and its metabolites have been measured in tissues 18 months after discontinuation of the drug. Other substances with similar characteristics include mercury, lithium, and chlorinated insecticides.

B. Modification of the Activity of the Drug-Metabolizing Enzymes

Some drugs detected in human milk are known modifiers of the activity of drug-metabolizing enzymes. Animal experiments have shown an increase in the *in vitro* ability to oxidize or reduce certain compounds following the intake via breast milk of phenobarbital (an inducing agent). Most likely, a similar phenomenon could occur in humans. An activation of drug metabolism results in a faster termination of drug action with a simultaneous effect on endogenous compounds such as steroids. An inhibition of the activity of this enzyme system prolongs the presence of unmodified drugs. The long-term consequences of the above are unknown, but should be evaluated.

C. Altered Response Genetically Determined

Minimal amounts of drugs which under normal conditions do not have any pharmacological effect may cause reactions in individuals with certain hereditary metabolic conditions. G6PD-deficient nursing infants may develop a hemolytic process when their mothers consume or receive substances known to cause such an effect (fava beans, sulfamethoxypyridazine).

D. Changes in Homeostasis Facilitating the Action of Other Agents

Anticoagulants cross the plasma-milk barrier in sufficient amounts to alter the infant's coagulation homeostasis. Any trauma, either surgical or mechanical, would then produce visible complications. A dramatic example exists of a 5-week-old infant, operated on for repair of inguinal hernia, who developed a huge hematoma at the site of surgery. He was a breast-fed infant whose mother was taking phenindione. His prothrombin time when measured was 50% of normal.

E. Interference with Normal Physiological Functions

The best-known example in this respect is the interference with thyroid function. Thiouracil is excreted into breast milk in high concentrations and in consequence the infant may present side effects due to the drug.

F. Hypersensitivity Reactions

Although not frequently seen, it is possible to develop hypersensitivity reactions to foreign compounds ingested through breast milk. A case report of such an event exists and refers to a mother and child receiving penicillin for the treatment of syphilis. The dose given to the infant was 10 units by injection while continuing on a regular breast-feeding schedule. Six hr after the mother received her high dose of penicillin, the child developed fever and localized reactions which persisted for 10 hr.

G. Adverse Drug Effects on a Normal Developmental Pattern

Several enzymic systems are not fully developed in the neonate, and mature at different rates until attaining adult levels. Small amounts of foreign compounds ingested with breast milk could interfere with such processes, for example, inhibitors causing the so-called breast-milk jaundice. The action in this case is on the maturation of the enzyme glucuronyl transferase. Another possible example is the induction of methemoglobinemia in infants of mothers who receive phenytoin and phenobarbital. This condition in newborns is attributed to a deficient diaphorase activity.

III. DRUGS EXCRETED INTO MILK

The theoretical considerations grouped under mechanisms and the respective examples found in the literature permit a more rational approach to the problem. From a practical point of view, present knowledge permits the division of substances excreted into breast milk into two main groups: those associated with undesirable side effects and those that are not.

A. Examples of Drugs Causing Adverse Effects

Some drugs which cause adverse effects are only of historical interest, such as bromides (causing drowsiness and rash) or the use of lead nipple shields and the occurrence of lead encephalitis in the infant.

On the other hand, environmental pollutants are a very modern concern and among others, two serious documented disasters were caused by them. In Turkey, in 1956, a type of severe porphyria occurred following the ingestion of hexachlorobenzene. It was also noted in small infants who ingested the fungicide via breast milk. In Japan, Minamata disease or mercury intoxication was produced by ingestion of contaminated fish in adults and breast milk in babies.

Signs of ergotism are developed by 90% of infants ingesting ergot alkaloids excreted in milk. Today, these compounds are present in a number of antimigraine preparations.

Narcotics are excreted in milk, and when ingested, may produce signs of addiction in the infant. This was well documented at the beginning of this century, and the recommended therapy was gradual weaning.

A commonly used preparation at the present time is oral contraceptives. Although there is little doubt that progestin-estrogen combinations appear in breast milk, the amount excreted and the biological activity of these steroids remain to be determined. Gynecomastia has been the only complication reported in a single case of a breast-fed infant whose mother received norethynodrel with mestranol. The long-term effects (physiological and/or behavioral) of oral contraceptives on breast-fed infants are not known, but should be looked for in a prospective study.

Anticoagulant therapy of the nursing mother and complications occurring in her infant have been discussed under mechanisms.

Early communications have stated that only small amounts of ethyl alcohol are present in breast milk. Nevertheless, a case of acute alcohol intoxication in an infant has been reported following the ingestion of a significant amount of port wine by the mother.

B. Examples of Drugs Not Known to Cause Serious Adverse Effects

The analgesics constitute the most commonly used group of drugs, and an on-going study regarding aspirin will be presented under experimental studies. Propoxyphene (Darvon®) passes into breast milk and has been found to be half of the amount present in the mother's plasma. No adverse reactions in infants have been detected following its occasional intake by the mother.

Antibiotics and chemotherapeutic agents do pass into breast milk in small concentrations. If absorbed, the levels in the blood of the infant would not be significant. Some theoretical complications have been analyzed under mechanisms, but presently no human documentation of their occurrence exists. Therefore, in general, they can be administered for maternal disease.

Among the anticonvulsant drugs, phenobarbital, which is excreted into breast milk, may modify the activity of the drug-metabolizing enzymes in the infant. Diphenylhydantoin also crosses into breast milk, and the appearance of undesirable effects in the breast-fed infant is controversial. There is only one case reported of methemoglobinemia in an infant whose mother was taking not only this drug but phenobarbital as well. Actual measurements of the drug in breast milk of women on continuous anticonvulsant therapy attest to the fact that the mammary gland has a limited ability to excrete this compound. A recent study (4) analyzed the transfer of diphenylhydantoin to the fetus and its passage into breast milk after delivery. The mothers received the drug (300 mg/day) and for 5 days postpartum therapeutic concentrations were attained in the mother's plasma (3.6 mg/ml), significant but decreasing amounts in the infant's plasma (2.6 to 0.6 mg/ml), and the concentrations in milk varied between 1.5 and 2.0 mg/ml. Only one patient on continuous therapy was restudied after 30 days, and she continued to transfer the drug into milk (1.3 mg/ml). Unfortunately, the concentration of diphenylhydantoin in the infant's plasma at that time is not recorded. Also, no mention is made of the time relationship between drug intake and sampling of milk and plasma.

No reports of toxic effects in the infant have been found regarding the use by the mother of antihistamines, fluoride, or lithium. Although lithium is excreted in breast milk at concentrations of one-fourth to one-third the maternal serum level, no toxic effect was observed in breast-fed infants (5).

The use of diuretics in the puerperal period is not uncommon. Although thiazide drugs have been measured in breast milk, no reports of toxic effects in infants have been found. Recently, the excretion of chlorothiazide (6) into breast milk in 11 nursing mothers was quantitated, but not the absorption by the infants since nursing was withheld during the study. The measurements showed that there was no significant transport of the drug into breast milk. By calculation, it was concluded that the amount of drug potentially transferred to the infant is very small. Therefore, the risk of a nursing infant acquiring toxic or even significant doses of the diuretic through breast milk is minimal.

Certain hormonal preparations—insulin, epinephrine, and corticotropin—are destroyed during their passage through the gastrointestinal tract, making their presence in breast milk totally unimportant. Although thyroxine remains biologically active after absorption, the amount excreted into breast milk is insignificant and therefore unlikely to produce adverse effects.

IV. EXPERIMENTAL STUDIES

Isolated measurement of a drug in breast milk is of little significance by itself and should be supplemented by the knowledge of the total quantity ingested and absorbed by the infant. For that purpose a research project was initiated at Children's Hospital in Buffalo, New York, in collaboration with Dr. L. K. Garrettson from the Department of Pediatrics and Dr. G. Levy from the Department of Biopharmaceutics of the School of Pharmacy. The objective was to measure the excretion of aspirin into milk after the administration of a single dose to the mother and to ascertain the amount of the drug absorbed by the infant. The subjects were selected from postpartum mothers in the maternity section of the hospital among those who expressed interest in breast feeding their infants. They volunteered after the study was explained in detail, first by the nurse in charge of the breast-feeding program and then by one of the participating physicians. The support of the family obstetrician and pediatrician was sought, and the protocol was discussed with them. The four mothers studied were college graduates residing in a suburban area and were enthusiastically committed to the project. They were admitted to the Clinical Research Center of the hospital for a stay of 48 hr, after the infants were at least 3 weeks old. Every effort was made to provide a pleasant environment, fathers could come at will, and mothers were allowed to move freely, except for nursing times and urine collections.

Neither mother or child had taken salicylate for a week prior to admission. After an overnight fast, the baby suckled until satiated and then both breasts were pumped to express all remaining milk. The mother was given sodium salicylate (20 mg/kg) in solution to facilitate absorption and resumed her regular diet 2 hr later. The infants did not receive any supplemental feeding. The total urinary outputs from mother and child were collected every 4 hr for 24 hr and at time of spontaneous voiding for another 24 hr. Before each feeding, a 10-ml sample of milk was put aside for analysis, the time accurately recorded, and the infants were weighed before and after each feeding.

In the organism, aspirin is first hydrolyzed to salicylic acid (the form given to our volunteers), which is then conjugated in part with glycine to form salicyluric acid and with glucuronic acid to form acyl and phenolic glucuronides. A small fraction is further hydrolyzed to gentisic acid. These compounds are then excreted by the kidneys in a combination of glomerular filtration and tubular secretion. Because approximately 98% of the drug and metabolites are excreted in the urine, the quantity recovered permits an assessment of the amount absorbed. The preliminary results obtained may be seen on Table 1.

TABLE 1. *Transfer of salicylate from nursing mothers to infants*

	Dose (mg)	Amount recovered (mg)		
		Mother	Infant	Total
1	1046	967	3.4	970.4
2	949	879	3.4	882.4
3	1380	1262	2.5	1264.5
4	1058	952	2.7	954.7

The total amount of salicylic acid (in mg) taken by the mother is shown, as well as the total amount recovered in urines. The recovery comprises salicylic acid and metabolites and is expressed in terms of salicylic acid. The amounts recovered from mother and child, respectively, are also specified. For the infants, considering that their body weights were approximately 3.4 kg, it is possible to estimate that they received in the vicinity of 1 mg of salicylic acid per kilogram body weight. Although the breakdown into the different metabolites is not yet available, it is interesting to note that salicylurates were found in the urine of the infants but not in the milk, indicating their ability to conjugate the drug with glycine.

The same results expressed as percent recovery are seen in Table 2.

TABLE 2. *Transfer of salicylate from nursing mothers to infants*

	Dose (mg.)	Amount Recovered (%)		
		Mother	Infant	Total
1	1046	92.45	0.33	92.78
2	949	92.58	0.36	92.94
3	1380	91.44	0.18	91.62
4	1058	89.98	0.25	90.23

Amounts recovered in the infants varied between 0.2 and 0.35% of the total amount. The total percentage recovered was slightly below the standard of the laboratory for young adult male volunteers. The study involved the administration of a single dose to the mother, and the results seem to indicate that it may be considered safe for the nursing infant. No data is available regarding chronic intake of the drug by the mother, although it could be assumed that under those circumstances the exposure of the infant to the compound would be rather significant. The pH of human milk was obtained as well around the clock, and proved to be quite stable. In contrast to animal findings, it is closer to the pH of plasma, and in the same lactating woman was recorded at approximately 7.2 to 7.3.

Diazepam, a frequently used tranquilizer, was reported originally as not being measurable in breast milk, but later studies differed reflecting improved laboratory techniques. This compound is frequently used in the postpartum period to relieve anxiety and nervousness in the mother, and therefore its possible transfer to the infant must be assessed. Such a study was designed by a team in Finland, and their preliminary results were recently published (7). Three postpartum mothers received the drug in doses of 10 mg, three times a day. On days 4 and 6 of the study, the concentrations of diazepam and of an active metabolite N-demethyl-diazepam in their plasma and milk, as well as in the plasma of the infants, was determined by gas chromatography.

Although there were differences in the concentrations of the drug measured on days 4 (491 ng/ml) and 6 (601 ng/ml) in the mother's plasma, these were considered not significant. The amounts present in breast milk (51 ng/ml and 78 ng/ml, respectively) were markedly lower than the corresponding values in plasma. In the infant, on day 4 the concentration in plasma of the metabolite (243 ng/ml) was much higher than the parent compound (172 ng/ml), reflecting that it is not only transferred in milk, but also that the neonate is able to demethylate the drug at an early stage. On day 6, both the parent compound (74 ng/ml) and the metabolite (31 ng/ml) were significantly lower than on day 4 (different from the observation in the mothers). The authors considered the possibility of a smaller intake of milk and therefore of drug, but quick calculations disproved this hypothesis. The onset of efficient elimination mechanisms in the infant seems to be the most logical explanation for this observation.

It is obvious that this drug given to the mother for her treatment is also able to produce a pharmacological effect in the infant, although the researchers did not see any visible reactions. Also, diazepam is conjugated with glucuronic acid and a possible competition with bilirubin could occur with the concomitant known complications.

V. PRACTICAL CONSIDERATIONS

1. Drugs not to be given nursing mothers:

Antimetabolites	Ergot
Most cathartics	Atropine
Radioactive drugs	Metronidazole (Flagyl®)
Anticoagulants	Dihydrotachysterol
Tetracycline	Thiouracil
Iodides	

2. Drugs to be given under supervision:

Oral contraceptives	Diuretics
Lithium carbonate	Nalidixic acid
Sulfonamide	Barbiturates
Reserpine	Diphenylhydantoin
Steroids	Cough medicine with codeine
Diazepam	

3. Drugs which are not contraindicated:

Occasional aspirin	Antidiarrheals
Most antibiotics	Epinephrine
Antihistamines	Insulin
Chlorothiazide	

This table is self-explanatory and can serve only as a guide.

VI. CONCLUSIONS

Breast milk meets the requirements of growing infants and is considered by many to be the ideal food for that age group. But two current events — environmental pollution and the widespread and indiscriminate use of drugs — may threaten its safety for the nursing infant. The proliferation of reports regarding the presence of drugs in breast milk and/or the appearance of side effects in the nursing infant indicates the awareness of the medical profession in this respect. These clinical observations have been followed by a few well-designed studies which should be extended to more of the most frequently used drugs. The forthcoming recommendations to nursing mothers would then be based on facts, and schedules might be designed that permit continuation of breast feeding while adjusting drug dosages.

ACKNOWLEDGMENT

The studies carried out in the authors' laboratories were supported in part by U.S. Public Health Service grants HDO4287, HDO6611, and RR-628 from the General Clinical Research Centers Program of the Division of Research Resources, National Institutes of Health.

REFERENCES

1. Rasmussen, F. (1966): *Studies on the Mammary Excretion and Absorption of Drugs.* Veterinaer-OG Landbohojskolen, Copenhagen.
2. Knowles, J. A. (1965): *J. Pediat.*, 66:1068.
3. Catz, C. S., and Giacoia, G. P. (1972): *Pediat. Clinic N. Am.*, 19:151.
4. Mirkin, B. L. (1971): *J. Pediat.*, 78:329.
5. Tunnessen, W. W., Jr., and Hertz, C. G. (1972): *J. Pediat.*, 81:804.
6. Werthmann, M. W., Jr., and Krees, S. V. (1972): *J. Pediat.*, 81:781.
7. Erkkola, R., and Kanto, J. (1972): *Lancet,* i, 1235.

Dietary Lipids and Postnatal Development
Raven Press, New York © 1973

Identification of Drugs and Drug Metabolites in Breast Milk by Gas Chromatography — Mass Spectrometry

M. G. Horning, J. Nowlin, P. Hickert, W. G. Stillwell, and R. M. Hill

Institute for Lipid Research and the Department of Pediatrics, Baylor College of Medicine, Houston, Texas 77025

I. INTRODUCTION

Recent studies indicate that the exposure of the fetus to drugs and drug metabolites during gestation is more extensive than is generally appreciated (1). Most pharmacologically active compounds administered to the gravid female cross the placenta and are stored in fetal tissues; the fetus is not protected by a "placental barrier" except with respect to macromolecules. Drugs administered to the mother during labor and delivery are also transferred to the fetus. After birth, the drugs and drug metabolites stored in neonatal tissues are metabolized and excreted; the disposal of drugs and drug metabolites may require several days to several weeks. Fortunately, many of the enzyme systems involved in the hydroxylation and conjugation of drugs are active in the human neonate from birth (1).

The exposure of the neonate to pharmacologically active compounds may continue for several months after birth as a result of breast feeding. Nearly all drugs taken by the mother are transferred to breast milk (2, 3). If the mother is on chronic drug therapy (epilepsy, arthritis, diabetes), the quantity of active drugs ingested daily by the neonate may approach therapeutic levels. Even a short exposure to a very active drug present in breast milk may affect the development of the neonate.

Very little quantitative data concerning the concentration of drugs in breast milk have been reported (3). This is probably due to the limited quantities of breast milk usually available, and to difficulties involved in measuring the low concentrations of drugs normally found in breast milk. Analytical procedures having high sensitivity and specificity of detection are required. Gas-phase analytical methods [gas chromatography (GC)

257

and gas chromatography-mass spectrometry computer techniques (GC-MS-COM)] are ideally suited for the study of this problem. In our laboratory, a simple and rapid procedure has been developed for the isolation of drugs and drug metabolites from colostrum and breast milk; quantification and identification are carried out by GC and GC-MS-COM procedures. The results of some preliminary studies are described in this chapter.

II. EXPERIMENTAL

A. Materials

Phenobarbital (5-ethyl-5-phenylbarbituric acid), secobarbital [5-allyl-5-(1-methylbutyl)-barbituric acid], caffeine (1,3,7-trimethylxanthine), diphenylhydantoin (5,5-diphenylhydantoin), morphine (7,8-dehydro-3,6-dihydroxy-4,5-epoxy-N-methylmorphinan), and the corresponding radioactive compounds were obtained from commercial suppliers. Valium® (7-chloro-1,3-dihydro-1-methyl-5-phenyl-2H-1,4-benzodiazepin-2-one) and ^{14}C-Valium® were a gift from Hoffmann-La Roche, Nutley, New Jersey.

B. Methods

1. Sample collection. Colostrum samples were collected from gravid females from 3 months prior to delivery to term and from mothers 1 to 3 days post delivery. Control breast milk samples were obtained from two mothers from 8 to 48 days after delivery. Random breast milk samples were also obtained from 4 to 29 days after delivery from a mother maintained on diphenylhydantoin and 2-ethyl-2-methylsuccinimide (ethosuximide, Zarontin®), and from a mother maintained on diphenylhydantoin alone.

All samples were stored at −14°C until analyzed.

2. Isolation of drugs from colostrum and breast milk. Samples of colostrum (0.5 to 1.0 ml) were transferred to a 15-ml screw-capped centrifuge tube and diluted to 5 ml with glass-distilled water. Four milliliters of ethyl acetate was added followed by 2.5 g of powdered anhydrous ammonium carbonate. The contents of the tube were thoroughly mixed by inversion and then allowed to stand for 5 min. The phases were separated by centrifugation for 15 min. After transferring the ethyl acetate (upper phase) to a 15-ml screw-capped centrifuge tube, the extraction was repeated by adding 2 ml of ethyl acetate to the ammonium carbonate solution, mixing thoroughly, and centrifuging for 15 min. The combined ethyl acetate extracts (approximately 6 ml) containing drugs and drug metabolites were evaporated with the aid of a nitrogen stream. The residue was dissolved in 0.5 ml of methanol for derivative formation and subsequent analyses by GC and GC-MS-COM. Isopropanol can be substituted for ethyl acetate.

It was necessary to modify this procedure before applying it to samples of breast milk because of the high concentration of free fatty acids present in most breast milk samples (Fig. 1). The ethyl acetate extract of drugs (and drug metabolites) isolated from breast milk as described above was evaporated (nitrogen stream). The residue was dissolved in 1 ml of ethanol-acetic acid-water (8:1:1) and extracted twice with 2-ml portions of hexane to remove fatty acids and neutral lipids. The ethanolic solution of drugs was then evaporated (nitrogen stream) and the residue dissolved in 0.5 ml of methanol and converted to derivatives. The fatty acids can also be removed by partitioning between methanol and isooctane as described previously (4).

The recovery of drugs from colostrum and breast milk was followed by adding trace quantities of radioactive drugs to the biological samples. The results are summarized in Table 1.

TABLE 1. *Recovery of radioactive drugs added to breast milk*

Drug	Ethyl acetate extract (%)	Isopropanol extract (%)	Hexane-washed ethanol (%)
Phenobarbital	98 ± 2	98 ± 2	79 ± 3
Caffeine	91 ± 2	93 ± 2	77 ± 3
Diphenylhydantoin	98 ± 2	98 ± 2	77 ± 4
Morphine	88 ± 3	98 ± 2	71 ± 4
Valium®	97 ± 2	94 ± 3	70 ± 3
Secobarbital	84 ± 2	98 ± 2	82 ± 3

3. Derivative Formation. Methylation followed by silylation usually provides derivatives with good gas chromatographic properties (5). An excess of ethereal diazomethane prepared from Diazald® (Aldrich Chemical Co.) was added to the methanolic solution (0.5 ml) of drugs. After standing at room temperature for 15 min, the excess ether, diazomethane, and methanol were removed (nitrogen stream) and the residue was dissolved in 10 μl of pyridine. *bis*-Trimethylsilyltrifluoroacetamide (BSTFA) (10 μl) and trimethylchlorosilane (TMCS) (5 μl) were added to the pyridine solution of methylated drugs. After heating at 60°C for 1 to 2 hr, an aliquot (2 to 5 μl) was used for analysis by GC or GC-MS.

4. Gas Chromatography-Mass Spectrometry. Gas chromatographic separations were carried out with 12 ft × 4 mm glass W columns containing

5% SE-30 on 80–100 mesh acid washed and silanized Gas Chrom P (6). Mass spectrometric studies were carried out with an LKB 9000 GC-MS-COM system (PDP-12 computer). The gas chromatographic column was a 9 ft × 3.4 mm glass coil with a 1% SE-30 packing. Mass spectra were recorded at 70 eV with an accelerating voltage of 3.5 kV and an ionizing current of 60 μA. The ion source temperature was 250°C. All GC (including GC-MS) separations were carried out by temperature programming from 90°C at 2°/min.

Single ion monitoring studies for diphenylhydantoin were carried out under isothermal conditions at 208°C with a multiplier setting of 11 and the exit and entrance slit at 1.2 mm and 0.3 mm, respectively. The ion m/e 180 (the base peak in the electron impact spectrum of diphenylhydantoin) was monitored.

III. RESULTS AND DISCUSSION

A rapid procedure for the extraction of neutral drugs with ethyl acetate or isopropanol from diluted breast milk (or colostrum) saturated with ammonium carbonate has been developed. The method is a modification

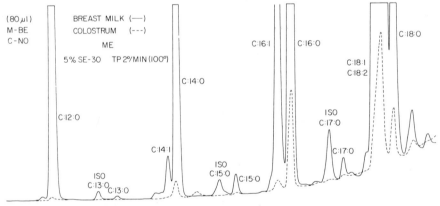

FIG. 1. Comparison of the GC analysis of human breast milk and colostrum. The gas chromatographic analysis of the ethyl acetate extract of breast milk is represented by the solid line and colostrum by the broken line. The equivalent of 80 μl of breast milk and 80 μl of colostrum was analyzed by GC. The compounds were separated as the methyl ester (ME) derivatives by temperature programming from 100°C at 2°/min on a 12-ft 5% SE-30 column. The peaks identified by GC-MS were C:12:0, lauric acid; C:14:1, myristoleic acid; C:14:0, myristic acid; C:16:1, palmitoleic acid; C:16:0, palmitic acid; C:18:1, oleic acid; C:18:2, linoleic acid; C:18:0, stearic acid; C:13:0, C:15:0, and C:17:0 are tridecanoic acid, pentadecanoic, and heptadecanoic acids, respectively; the corresponding C:13:0, C:15:0, and C:17:0 iso acids were also present.

of the isopropanol-potassium carbonate salting-out procedure developed for urine (7) and plasma (4). Ammonium carbonate was substituted for potassium carbonate; this results in a decrease in the temperature of the aqueous phase instead of the marked increase in temperature that accompanies saturation with potassium carbonate.

No difficulties have been encountered with the analyses of colostrum. However, breast milk has a much higher content of free fatty acids and related lipids than colostrum, and ME-TMS derivatives of these compounds interfere with the qualitative and quantitative analyses of drugs in human breast milk. A gas chromatographic analysis of an extract of colostrum (80 μl) and of breast milk (80 μl) are compared in Fig. 1. The concentration of free fatty acids (analyzed as the methyl esters) in the breast milk sample is approximately 20 times that of the colostrum sample. Odd-numbered straight-chain and branched-chain fatty acids, as well as lauric, myristic, myristoleic, palmitic, palmitoleic, stearic, oleic, and linoleic acids, were identified in breast milk. Monoglycerides, including monocaprin through monostearin, were eluted from the column at higher temperatures.

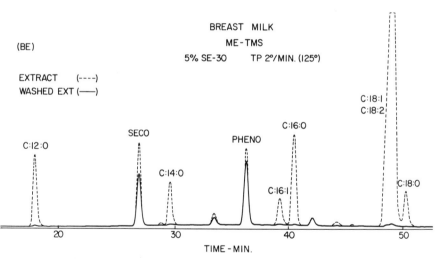

FIG. 2. Analysis of human breast milk. The gas chromatographic analysis of the ethyl acetate extract of breast milk to which secobarbital and phenobarbital had been added is represented by the broken line; the analysis of the hexane-washed extract is represented by the solid line. Small losses of secobarbital and phenobarbital were observed in the washed extract. The conditions of separation of the methylated derivatives were the same as those described for Fig. 1 with temperature programming from 125°C. The peaks identified by GC-MS were SECO, secobarbital; PHENO, phenobarbital; C:12:0, lauric acid; C:14:0, myristic acid; C:16:1, palmitoleic acid; C:16:0, palmitic acid; C:18:1, oleic acid; C:18:2, linoleic acid; and C:18:0, stearic acid.

Interfering fatty acids can be removed from breast milk samples by extraction with hexane or isooctane as described above. A comparison of a gas chromatographic analysis of the initial ethyl acetate (or isopropanol) extract with the hexane-washed extract of breast milk is shown in Fig. 2; secobarbital and phenobarbital had been added to the breast milk sample. Most of the fatty acids present in the initial ethyl acetate extract have been

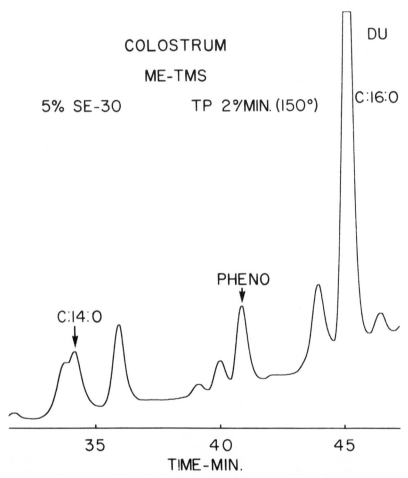

FIG. 3. Gas chromatographic analysis of the ethyl acetate extract of human colostrum, collected 3 days after delivery from a mother maintained on phenobarbital (195 mg/day). The compounds were separated as the methylated derivatives, using conditions described in Fig. 1 with temperature programming from 150°C. The peaks identified by GC-MS were C:14:0, myristic acid; C:16:0, palmitic acid; and PHENO, phenobarbital.

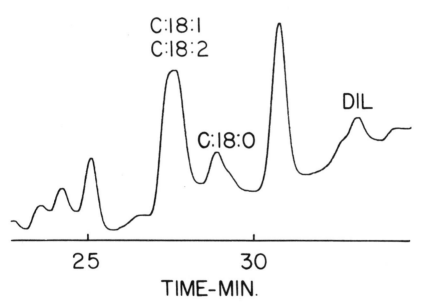

BREAST MILK

ME-TMS

5% SE-30   TP 2°/MIN. (190°)

BU-2 HR.

100 mg DIL

C:18:1
C:18:2

DIL

C:18:0

25   30

TIME-MIN.

FIG. 4. Gas chromatographic analysis of the isopropanol extract of human breast milk collected 19 days after delivery. The mother was maintained on 300 mg of diphenylhydantoin (DIL) per day. The compounds were separated as the methylated derivatives using the conditions described in Fig. 1 with temperature programming from 190°C. The peaks identified by GC-MS are C:18:1, oleic acid; C:18:2, linoleic acid; C:18:0, stearic acid; and DIL, diphenylhydantoin.

removed from the washed extract by hexane extraction. As can be seen from the gas chromatographic tracing, a small loss of phenobarbital and secobarbital occurred as a result of the hexane treatment.

The recovery of phenobarbital, caffeine, diphenylhydantoin, morphine, Valium®, and secobarbital was followed by adding trace amounts (50 ng to 2 μg) of the radioactive drugs to colostrum and breast milk (Table 1). The recovery was excellent (93 to 99%) for all of the drugs when isopropanol was employed. When ethyl acetate was substituted for isopropanol, the recoveries varied from 84 to 98%. The separation of the two phases, however, occurs more readily when ethyl acetate is used as the extractant. If labeled drugs are not available, the recovery can be followed on a mass basis using standard gas chromatographic procedures (8).

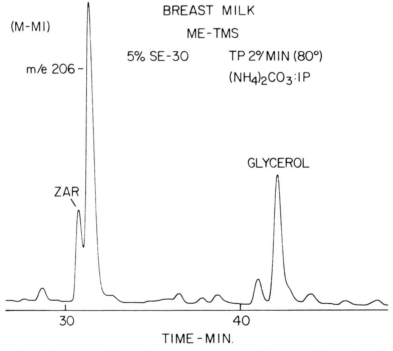

FIG. 5. Gas chromatographic analysis of the isopropanol extract of human breast milk collected 36 days after delivery. The mother was maintained on 1250 mg of 2-ethyl-2-methylsuccinimide (Zarontin®) per day. The compounds were separated as the methylated and silylated derivatives using the conditions described in Fig. 1 with temperature programming from 80°C. The peaks identified by GC-MS were ZAR, Zarontin®; glycerol; and an unknown, *m/e* of 206 (ethylene glycol).

In a separate series of experiments, unlabeled drug was added in addition to labeled drug to demonstrate that the intact drug as well as radioactivity had been recovered and that the drugs could be converted to derivatives for analysis by GC and GC-MS (Fig. 2).

Although it was possible to carry out quantitative analyses for drugs present in some of the isolated samples by use of gas chromatographic procedures (hydrogen flame detector), most drugs in breast milk could not be estimated accurately by this method. Figures 3–6 illustrate some of the difficulties encountered. When the drug was present in relatively large quantities and formed a well-defined GC peak that was homogeneous by mass spectrometric analysis, gas chromatographic analysis was satisfactory. For example, the peak labeled PHENO (phenobarbital) in the colostrum sample illustrated in Fig. 3 was homogeneous by mass spectrometry and could be measured accurately by GC methods.

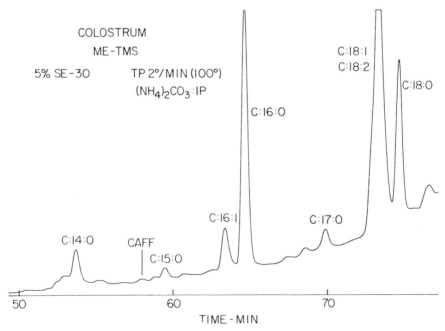

FIG. 6. Gas chromatographic analysis of the isopropanol extract of human colostrum collected 2 months before delivery. Two cups of coffee had been ingested 1 hr before collection of the sample. The compounds were separated as the methylated derivatives using the conditions described in Fig. 1. The peaks identified by GC-MS were CAFF, caffeine; C:14:0, myristic acid; C:15:0, pentadecanoic acid; C:16:1, palmitoleic acid; C:16:0, palmitic acid; C:17:0, heptadecanoic acid; C:18:1, oleic acid; Ca; C:18:2, linoleic acid; and C:18:0, stearic acid.

However, in Fig. 4, the peak identified as DIL (diphenylhydantoin) was not well separated from the adjacent peak, and precise measurements would be difficult to obtain by gas chromatographic procedures. The symmetrical peak identified as ZAR (2-methyl-2-ethylsuccinimide) in Fig. 5 could be considered satisfactory for quantitative analysis. However, analysis by mass spectrometry indicated that the peak was not homogeneous.

The peak identified as CAFF (caffeine) in Fig. 6 was too small to measure accurately, and the sample of colostrum provided only enough material for a single GC and a GC-MS analysis. In this sample, however, it was possible to identify caffeine by MS because of characteristic ions of high intensity present in the mass spectrum at 109 and 194 amu.

From these examples, it is clear that both greater sensitivity and specificity of detection are required for quantitative analyses of drugs in breast milk. The sensitivity of the mass spectrometer operated in the single or multiple ion mode (9–11) is greater than most hydrogen flame detectors. In addition, it is usually possible to measure an ion that is characteristic, if not unique, for the compound under investigation, thus providing specificity as well in the measurements.

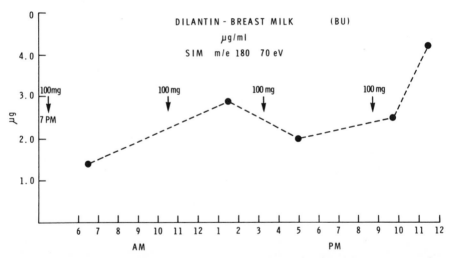

FIG. 7. Analysis of isopropanol extract of human breast milk by single ion monitoring using an LKB-9000 mass spectrometer; the ion *m/e* 180 was monitored. Samples were obtained 19 days after delivery at 6:30 A.M., 1:30 P.M., 5:00 P.M., 9:45 P.M., and 11:30 P.M. Diphenylhydantoin (dilantin) (100 mg) had been ingested at 10:30 A.M., 3:15 P.M., and 8:45 P.M., and at 7:00 P.M. on the previous day.

Analyses were carried out on extracts of breast milk and colostrum samples using the technique of single ion monitoring. The results of one of the studies are shown in Fig. 7 and 8. A series of breast milk samples were obtained at 6:30 A.M., 1:30 P.M., 5:00 P.M., 9:45 P.M., and 11:30 P.M. from a mother maintained on diphenylhydantoin (300 mg/day as three 100 mg doses per day). The concentration of diphenylhydantoin, measured by selective ion monitoring, varied from 1.4 μg/ml to 4.2 μg/ml. The concentration of drug was lowest in the 6:30 A.M. sample; this was 11.5 hr after ingestion of the last dose of 100 mg of diphenylhydantoin (Fig. 7). The highest concentration was observed at 11:30 P.M., after ingestion of the third 100-mg dose. It is interesting that the concentration of drug in breast milk was higher

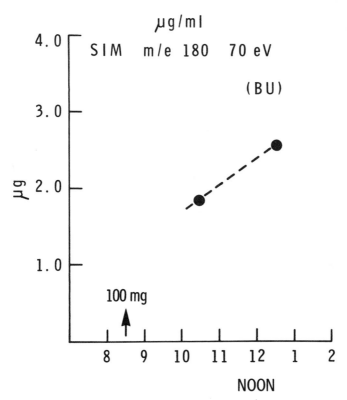

FIG. 8. Analysis of isopropanol extract of human breast milk by single ion monitoring using the LKB-9000 mass spectrometer; the ion *m/e* 180 was monitored. Samples were obtained 38 days after delivery at 10:30 A.M. and 12:30 P.M. following ingestion at 8:30 A.M. of 100 mg of diphenylhydantoin.

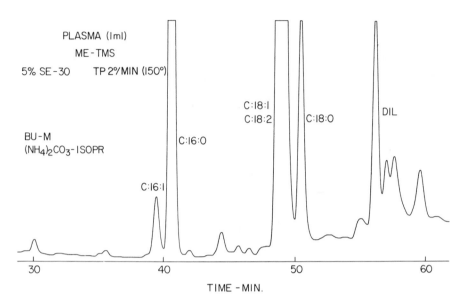

FIG. 9. Gas chromatographic analysis of the isopropanol extract of plasma collected at delivery from the mother maintained on 300 mg of diphenylhydantoin per day. The compounds were separated as the methylated derivatives using the conditions described in Fig. 1 with temperature programming from 150°C. The peaks identified by GC-MS were DIL, diphenylhydantoin; C:16:1, palmitoleic acid; C:16:0, palmitic acid; C:18:1, oleic acid; C:18:2, linoleic acid; and C:18:0, stearic acid.

at 2.75 hr (11:30 P.M.) than at 1 hr (9:45 P.M) after ingestion of the drug. (The gas chromatographic analysis of the breast milk sample obtained at 11:30 P.M. is shown in Fig. 4). This slow rise in levels of drug in breast milk was observed in samples collected at a later date (Fig. 8) from the same patient. The concentration of diphenylhydantoin was higher at 4 hr (12:30 P.M.) than at 2 hr (10:30 A.M.) after ingestion of the drug.

The concentration of drugs in plasma can also be determined through use of the isopropanol-ammonium carbonate extraction procedure. Figure 9 shows the gas chromatographic analysis of a plasma extract, from blood obtained at the time of delivery, for the mother who later provided the breast milk samples.

IV. CONCLUSION

Drugs administered to nursing mothers will be found in colostrum and breast milk. The diffusion model developed by Rasmussen (2) indicates that the concentration may be either greater or less than the concentration in plasma, depending on the degree of ionization and lipid solubility of the drug. Situations involving the chronic use of anticonvulsants and phenobarbital are of particular interest. Diphenylhydantoin, for example, is present in breast milk when administered to mothers. The exposure of the infant to drugs, which may have occurred throughout pregnancy, is therefore continued through the neonatal period.

ACKNOWLEDGMENT

This work was supported by grant GM-16216 of the National Institute of General Medical Sciences.

REFERENCES

1. Horning, M. G., Stratton, C., Nowlin, J., Wilson, A., Horning, E. C., and Hill, R. M. (1972): In: *Fetal Pharmacology,* edited by L. O. Boréus. Raven Press, New York.
2. Rasmussen, F. (1966): *Studies on the Mammary Excretion and Absorption of Drugs.* C. Fr. Mortensen, Copenhagen.
3. Knowles, J. A. (1965): *J. Pediat.,* 66:1068.
4. Horning, M. G., Boucher, E. A., Stafford, M., and Horning, E. C. (1972): *Clin. Chim. Acta,* 37:381.
5. Harvey, D. J., Glazener, L., Stratton, C., Nowlin, J., Hill, R. M., and Horning, M. G. (1972): *Res. Comm. Chem. Path. Pharmacol.,* 3:557.
6. Horning, E. C., Vanden Heuvel, W. J. A., and Creech, B. G. (1963): In: *Methods of Biochemical Analysis,* Vol XI, edited by D. Glick. Interscience, New York.
7. Bastos, M. L., Kananen, G. E., Young, R. M., Monforte, J. R., and Sunshine, I. (1970): *Clin. Chem.,* 16:931.
8. Horning, M. G., Chambaz, E. C., Brooks, C. J., Moss, A. M., Boucher, E. A., Horning, E. C., and Hill, R. M. (1969): *Anal. Biochem.,* 31:512.
9. Brooks, C. J. W., and Middleditch, B. S. (1972): *Clin. Chim. Acta,* 34:145.
10. Hammar, C. G., Holmstedt, B., and Ryhage, R. (1968): *Anal. Biochem.,* 25:532.
11. Sweeley, C. C., Elliott, W. H., Fries, I., and Ryhage, R. (1966): *Anal. Chem.,* 38:1549.

INDEX